Creature Comforts in Space

Designing Enjoyment & Sustainability for Off-World Living

Ken, it's all about the stories!

[illegible]

Praise for "CREATURE COMFORTS IN SPACE"

"Creature Comforts in Space is a key to unlocking the monumental opportunity of Large-Scale Space Migration (LSSM). If a substantial number of people and industries relocate off-planet, it will be incredibly beneficial to humanity, the Earth, and life itself. However, that is only going to happen if living in the rest of the solar ecosystem is comfortable for the vast majority of those who "boldly go." Creature Comforts tells us how to make that happen!"

– Frank White, space philosopher, author of "The Overview Effect"

"Outer space design pioneer Samuel Coniglio has been pushing the boundaries of creating ever better quality experiences for living off-world. This book pulls it all together as an invaluable resource to help move us all forward."

– John Spencer, Outer Space Architect, Founder of the Space Tourism Society

"Samuel Coniglio's Creature Comforts catalogs the things that will make the future of space more human. Going beyond pure function, it takes into account adaptation, purpose, pleasure, and sustainability. This book lays the foundation of how to properly practice space architecture so it welcomes the whole of humanity."

– Phnam Bagley, space architect and industrial designer at Nonfiction

"Designing aircraft interiors is not much different than designing spacecraft interiors. It is a balance of functionality and beauty, while keeping in mind the limitations of weight, space and resources. Samuel's book masterfully finds that balance. He states the design limitations of outer space, yet envisions domestic items like a kitchen that can work in such a strange place."

– Marie-Lise Baron Design, founder of THE BARON INSTITUTE,
International Design Institute of Luxury & Private Jets

Designing Enjoyment & Sustainability for Off-World Living

Samuel M. Coniglio, IV

Foreword by Rick Tumlinson
(co-founder, Space Frontier Foundation)

For more information, email: contact@retro-futurist.com

First Edition
Published in 2024

ISBN: 979-8-218-24640-2 (Paperback)
ISBN: 979-8-218-24639-6 (eBook)

Subjects: SCIENCE: Space Science/Space Exploration, SCIENCE: Space Science/General, SCIENCE: Philosophy & Social Aspects, ARCHITECTURE: Sustainability and Green Design, Industrial Design, Space Architecture

Audience: Architects, artists, designers, industrial designers, interior designers, space enthusiasts, space entrepreneurs

Keywords: aerospace, architecture, creature comforts, domesticity, engineering, ergonomics, extreme environments, futurism, homesteading, human factors, industrial design, interior design, NASA, off-world living, outer space, product design, mental health, space architecture, space history, space life sciences, space lifestyle, space settlement, space travel, space exploration, space station, sustainability, sustainable design, well-being

Cover design by S.N. Jacobson
Front graphic design by Sarah Dungan
Zero gravity cocktail glass and drinking cup 3D CAD designs by Nick Donaldson
Cover title font: Planny (myfonts.com), Cover text font: Blueprint (fonts.com), Manuscript font: Arial

The concepts described in this book are design exercises to educate and inspire designers and the design-curious. There is no guarantee that they will work in real life. These are starting points for designers and engineers to build a sustainable and comfortable off-world lifestyle. If you would like to collaborate with the author in building the real thing, please contact the author at https://retro-futurist.com/.

The author's team have made all reasonable attempts to credit the creators of the photos, and the persons in them, throughout the book. If a person has not been properly credited, please contact the author at https://retro-futurist.com/.

Printed in the United States of America

https://retro-futurist.com/

This book is dedicated to Sebastian Apollo Coniglio,
whom I hope will live off-world one day.

This book is also dedicated to Katherine Becvar,
my wife and partner in crime,
who inspired me to finally embark upon this project,
gave sage advice,
and kept me sane through it all.

Contents

Foreword
by Rick Tumlinson

As I write this, it's over 39° Celsius (103° Fahrenheit) outside. It might not kill me, but it's not comfortable. Regardless, I am here in my little bubble of an apartment with all its amenities. During COVID, I was here for days at a time without venturing into the environment of the planet at all. I was fine. I had air, water, food, and a regulated temperature—all I needed to survive. I had communications with other humans, and all the entertainment I wanted was wired or beamed in. I had my food delivered. I had my weights, and although I was lazier than I might admit, I was able to stay in shape. Heck, I added some muscle by the end of it all. Frankly, I could have been anywhere. I could have even been on the Moon or Mars. Or, more realistically, given the challenge of radiation, under the Moon or Mars. I would be fine. You would be fine. For the most part, we were fine.

We watch and read sci-fi and future fiction and don't realize that between the action stuff, when not popping around the galaxy at warp speed, fighting alien invaders, or exploring "strange new worlds," the characters we are watching go home, strip off their skin suits and live in apartments, tiny homes or modules that are essentially the same as ours. The locations are different. The materials are different. The climate or environment is different. But the day-to-day needs and routines of life of every human being in our post-industrial so-called modern societies has reached a point of standardized stability that is probably not going to change that much for the next thousand years, be it here on Earth or floating in a bubble community above the boiling surface of Venus.

Yes, within the matrix of those lives, things will change. The shows currently viewed on our flat screens will be beamed into our brains, and probably at some point, we will "live" them in full flesh-feeling multi-dimensional media. The food we eat will be hyper-nutritious vegan, incredibly-realistic meat-imitating protein, maybe even composed on the spot by Star Trek-like maker machines. We will plug into exercise machines that tense and release our muscles a million times a minute to keep us fit. And we will drift to sleep with the help of machine-induced melodies tuned to take our tired minds into places warm and wonderful.

And yet, we will still live in little bubbles, with other little bubble lovers and friends and families who each live in their own little bubbles, all living in community bubbles of life, each of us paxt of an even bigger bubble of expanding life and humanity, growing outwards from the MotherWorld who gave us life to begin with.

Be it tomorrow or a thousand years from now, much of what happens in our day-to-day lives will be recognizable to you and me and any of our ancestors from a thousand years ago. There will be wonders! Oh, such wonders! Be it electric can openers or the cans themselves, someone somewhere on the human timeline will be amazed. Be it the talking head on my TV screen or the warming touch of an invisible hand on my body that feels so real it makes me tingle as if real, my sensory interaction with my environment will evolve but stay the same. This is because I, as a human, am the same.

That is the point. Until the AIs show up and sweep us into the robotic vacuum cleaner of SkyNet's net, I, the human in the loop, will remain. You will remain. We will remain. And as humans, we will still, between our grand adventures in exploration, seek the solace of our private little bubbles, our homes, our apartments, our cubicles, cubby holes, and crew quarters. There, in our own private moments, we will turn on the light or plug in the data cord and let our minds wander to yet other worlds, real, unreal, made up, or unmade, as that is what we humans do.

Samuel's book is not a book about space. It is a book about people. It is not about the technology of space. It is about the technology of being human. It is not about the exploration of the universe. It is about the exploration of the tiny universe of our own human experiences and the comforts that make all of it worth doing.

I want to dance in the 3D multi-axis freedom of zero gravity, even though I know my inner ear will be screaming until it's trained, disconnected, or drugged into obeying my desire. I want to make love to my lover without the worries of falling off the bed—because the room we are in IS a bed, even knowing I will need straps and grabbing points because without them there is no leverage to give my love the full force of my passion. I love the idea of diving Up into a floating pool of water, even knowing that unless it is blown off my face when I come out the other side I will drown, as there is no gravity to pull it away from my nose.

These are the details. The realities of the reality of what it will really be like to be in space. But they are essential to consider. And thank the gods, someone like Samuel is thinking about them now. Look, someday the space programs of Russia, the U.S. and its allies, and China will be thanked for getting us out there. They will also be thanked for developing the many basic tools and tech needed to survive. But it has never been their job or role to help people Live in space. Live, as in enjoying themselves. The human spaceflight programs of all the world's governments were designed for scientific and strategic reasons. The psychology and engineering of comfort have only been implemented insofar as it supports the missions of the employees those governments have and are sending to space, Period. The rest is up to us.

Every add-on, be it the color paint in the spaceship, the way the toilet flushes, the smells it releases (or does not) into the hab module, the sleeping quarters, and the tastes and textures of the food, are all mere calculations designed to assure the maximum productivity, longevity, and psychological well-being of a crew who has a job to do.

The word “fun” has not, nor will ever appear in government design specifications. Pleasure does not appear in the engineering plans for a spaceship. Happiness, satisfaction, wonder, and delight are not words to be found in contractors’ paperwork.

Yet they are what it is all about. What life is all about. And yes, even what work itself is really, at its core, all about. No matter when in history, where in geography (or beyond geography), humans are people, and people want to have fun.

So as Samuel lays out the nits and grits of daily life in what our shared mentor Dr. Gerard K. O’Neill called the High Frontier, I hope it helps you realize that it is a human frontier. Gerry knew this. I recall working on the reprint of the book itself, the core tome around which all of us who work to open space orbit, The High Frontier. He brought us back to the human aspect as we went through variations on a possible cover. Rather than magnificent images of massive rotating habitats surrounded by cool spaceships, he insisted on a scene of a family having fun on a beach. Yes, in the background, you could see the curving vista of the incredible spherical bubble in which they lived, but it was the people that mattered to him.

That, in the end, is what it is all about. People. I speak in my writings of the Purpose of People. I believe it to be grand, exciting, inspiring, and something worth fighting for. Yet none of it is worth anything—unless it is fun. We go out there to “these and all the other things,” as Kennedy said. And among those things is to have a good time!

ENJOY!

–Rick Tumlinson
Co-Founder, Space Frontier Foundation
July 2023

(**Editor’s Note:** Rick is being modest. Here are some of the organizations he’s helped found: Deep Space Industries, Earthlight Foundation, LunaCorp, MIRCorp, New Worlds Institute, Orbital Outfitters, Space Fund, X PRIZE (founding trustee), and many others…)

Introduction

After 50 years of space exploration, off-world living is still like going on an extreme camping trip. Living out there is uncomfortable at best, and deadly at worst. It is a hostile environment that will kill you if you make a mistake. You have to bring all the supplies you need to stay alive.

Nevertheless, space travel is becoming more common. The Karman Line, the officially recognized dividing line between the Earth's atmosphere and the beginning of space, is 100 kilometers from sea level.[1] As of 2020, over 500 astronauts from almost 40 countries have traveled to space.[2]

The hard part has been done. We know how to travel to space. The space industry is working diligently on getting the cost down by increasing launches with reusable rockets. Once you get into orbit, though, day-to-day living is still a struggle.

This book is intended to inspire readers who are interested in design, who might be new to the space industry, and who want to know the challenges of creating an off-world habitat. I hope to give you a designer's perspective on improving comfort and fun in key areas of a future space station. Whether you are an architect, engineer, entrepreneur, policy maker, or just an avid space enthusiast, this book will help you understand the challenges of creating a safer, more comfortable "Space Lifestyle." Perhaps, it will make you reconsider many of the things we take for granted here on Earth today, and maybe even inspire you to change at least one wasteful habit into something more sustainable.

Why this Book Now?

NASA is no longer the only customer for space. Private companies like AXIOM Space, Blue Origin, and SpaceX have the capability of building viable space stations for private and government use. Many other private companies are developing new business opportunities which will create a diversified off-world environment.

---

1. *"100km Altitude Boundary for Astronautics | World Air Sports Federation," August 1, 2017, https://www.fai.org/page/icare-boundary.*
2. *Thomas Roberts, "International Astronaut Database," Aerospace Security, July 5, 2022, https://aerospace.csis.org/data/international-astronaut-database/.*

I have been researching space tourism as an industry since 1996. For many years, with the exception of a few rare pioneering space trips by the occasional wealthy traveler, the industry was not yet ready for prime time. Starting in 2021, when over a dozen private space travelers flew on non-NASA spacecraft for the first time, it became obvious to me that now is the time to start making the trip as comfortable and safe as possible for regular people, not just highly trained astronauts. Lowering the cost of space access and improving comfort during travel means more opportunity and accessibility to more people.

You may feel frustrated that we have not built these "cities in the stars" yet. You are not alone. For years, I listened intently to the ideas presented by space experts at conferences run by the American Institute of Aeronautics and Astronautics (AIAA), National Space Society, Space Frontier Foundation, and others. One issue that irked me at these space conferences was the lack of research on the human factor. All these very smart people seemed to ignore the obvious: how does a person survive for weeks or months on a space station, a lunar base, or a Mars hotel? *Cabin fever is a thing.* People will go stir crazy if they have to eat the same food and do the same routine every single day for months. One example of this was the so-called "rebellion" on NASA's Skylab space station back in 1974.[3] Onboard, a very overworked astronaut crew had every minute of their lives scheduled with very few breaks or downtime. They finally just turned off the comms link to Ground Control so they could have a breather. Mental health is a big issue in extreme environments and must be taken seriously. *Airlocks can be opened very easily…*

Creature Comforts and You

This book tackles the issue of creature comforts: those hard-to-describe things that make life more enjoyable, have mental health benefits, and are fun! Creature comforts include the simple things in life: having a cup of tea, taking a shower, being able to dance, or just living like a normal human being…while traveling in a place that is not safe for humans. In my years of research, I discovered that many of these creature comforts are interconnected with sustainability. This means water and air recycling, power generation, and other elements of self-sufficiency that can reduce the criticality of constant resupply shuttles from Earth.

Why do you need creature comforts? They are the little things: the habits of life that give it consistency and make you feel good. Feeling good improves morale, improves response time during emergencies, and otherwise makes you relaxed and more creative. Psychologists would agree that personal comforts enhance happiness, and if you can bring some (or adapt some) for your long space journey, it will make the trip more bearable.

---

3. L. A. Rockoff, R. F. Raasch, and R. L. Peercy, "Space Station Crew Safety Alternatives Study. Volume 3: Safety Impact of Human Factors," June 1, 1985, https://ntrs.nasa.gov/citations/19850021672.

What exactly are creature comforts? The answer is extremely subjective. Below is the result of a (very) informal survey I did with friends on Facebook. It seems that for each person you ask, you will get a slightly different answer. Here is a list of their most common answers.

What are Creature Comforts?
(A very unscientific survey from friends)

- Cold beer/ cocktails
- Eye drops
- Fluffy towels
- Fresh baked cookies
- Fresh coffee/ espresso/tea/ hot chocolate
- Hot meals/ Fresh food
- Heating/air conditioning
- Hot showers/ sauna
- Indirect lighting
- Pan fried bacon
- Plants/gardens/ farms
- Silence
- Sweat pants
- Toilets/bidet
- Washing hands
- Weight room

Example Creature Comforts. Source: The author.

Throughout this book, you will find sidebars that address each comfort in detail, located in relevant chapters.

What is Your Creature Comfort?

The list above is far from complete. As more humans begin to live off-world for long periods of time, other personal needs, habits, and interests will need to be addressed in future space station and planetary settlement designs. What are your top three creature comforts, and how would you adapt them to this strange new world?

For me, one of them is having a nice hot cup of tea while reading a book on a quiet morning. Since becoming a parent, such a simple luxury is as rare as hen's teeth. I have to negotiate with the wife and child to find those moments. Setting expectations with folks around you is extremely important, especially since they have their own needs or interests that might clash with yours. How might the needs and comforts of your space-faring companions interact with yours? How might that need to be accommodated so everyone's having the best time without sacrificing too much?

Design Constraints

This book takes a candid look at the dangers of space travel and seeks to turn them into design constraints for new concepts for living better off-world, and maybe improving life on Earth. Design constraints are limitations on a design project, whether it is self imposed or out of your control. They help keep the project focused. Example constraints include budget or time limits, or customer requirements. For this book, the design constraints are the extreme environment of space, and the practical limits of what one can launch into space with current rocket technology. The designs in this book are conceptual and deliberately kept at a high level, to hopefully inspire you to take them to the next level.

Design Analogies

The closest Earthbound design analogy I could find is that of a homestead: life on a farm or a remote cabin in the woods where you have to "live off the land" and be very efficient with resources. I also discovered that my off-the-grid adventures at Burning Man, which is an experimental city in the Black Rock Desert in northern Nevada, USA, has relevance to space travel because of its survival camping aspects. Learning to adapt to an unknown and harsh environment has parallels to space travel. This idea is not so far-fetched: survival camping is similar to military survival training, working in a remote Antarctic research station, or doing research in an underwater base.

Why am I Writing this Book?

I am compelled to write this book because someone has to. Of the many amazing space science communicators out there, very few dive deep into topics related to space architecture and the challenges of day-to-day living in an extreme environment.

My background is rather unusual. I am a generalist surrounded by specialists. With a talent of grabbing ideas from industries that have nothing to do with each other, I mix-and-match them to find new perspectives that can lead to new concepts. Some people call that a futurist. I prefer to call myself a retro-futurist: because I look at the past as well as the present and reinvent visions for the future. The following is a summary of my wacky life.

In the 1990's, fresh out of college, I worked for McDonnell Douglas at the Kennedy Space Center, in Florida. I was a technical writer documenting payload processing systems for both the U.S. Space Shuttle and for the International Space Station (ISS), which at the time was still under development. This was my first introduction to NASA and the space industry.

During my time at that company, I was also given the opportunity to work on a special side project, funded by the U.S. Department of Defense (DoD). It was called a Delta Clipper Experimental (DC-X).[4] Designed and built on a shoestring budget of $90 million (which is a drop in the bucket for the DoD), it was a project in which various teams across the company were able to collaborate and build a fully working experimental rocket in 18 months.[5] This was the world's first fully reusable rocket ship test bed. McDonnell Douglas had just enough funding to fly it once, and it survived to fly eleven

---

4. *Mark Wade, "DC-X," Encyclopedia, Encyclopedia Astronautica, December 28, 2012, https://web.archive.org/web/20121228125150/http://www.astronautix.com/lvs/dcx.htm.*

5. *G. Harry Stine, Halfway to Anywhere: Achieving America's Destiny in Space, 1st Edition (New York, New York: M. Evans and Company, Inc., 1996).*

more times because of the efforts of space activists. This was when I discovered the Space Frontier Foundation, a renegade band of space entrepreneurs, who were obsessed with finding other ways to fly into space without having to be dependent upon the 800 pound gorilla known as NASA.[6] This is where I learned of alternative space projects outside of NASA.

Ironically, the DC-X project finally ended up at the very place the space activists were trying to avoid: NASA. Fortunately, the Administrator at the time, Daniel Goldin, understood the long term vision the DC-X represented, and took it on as a role model for his new "Faster, Better, Cheaper" mantra that he was pushing on the traditional NASA contractors.[7] Renamed the DC-X/A, the rocket proved itself as a reliable test bed for new technologies. It flew a total of 12 times, with the final flight a victim of human error (someone forgot to install a critical bolt in the landing gear), and the rocket took its place in history.

Not finding a job with the space entrepreneurs (they only wanted engineers), and tired of the NASA bureaucracy, I left the space industry and became a freelancer and traveled the USA. I stayed deeply involved with many space organizations such as the Space Frontier Foundation, National Space Society, Planetary Society, American Institute of Aeronautics & Astronautics (AIAA), and others. This is where I learned many aspects of the space industry.

Since 1999, I have been involved with the Space Tourism Society.[8] It is a nonprofit organization based in Los Angeles, California, that researches and promotes the potential of a space tourism industry. During my tenure as Vice President, I helped to organize events and designed several concepts that would help improve passenger living in future orbital cruise ships or super yachts. We like using the cruise ship analogy for space travel because they're both basically floating in a harsh environment far away from civilization. I will give more details of some of these inventions in upcoming chapters.

One of my inventions, the Zero Gravity Cocktail Glass, was developed into a 3D printed prototype and marketed as part of a startup I co-founded called Cosmic Lifestyle Corporation, in 2014. The glass received design awards and international acclaim, but the business was a failure. Anecdote: the story I was told was that our friends at MADE IN SPACE (Now Redwire) came really close to getting the glass produced on one of their new 3D printers onboard the ISS, except that NASA Public Affairs Office got wind of it and said "NO WAY." Also, I learned a lot about how NOT to run a business, as well as how fluid flows in microgravity. More details about the glass are in Chapter 2.

Participating in two other space projects gave me insight into communicating, networking, and promoting space travel beyond just engineering and science. First, in 2004, I joined the Ansari XPRIZE TV Crew as a photographer, and spent two weeks in Mojave, California documenting the historic flights of Space Ship One. The Ansari X PRIZE competition granted the winner $10 million for successfully flying a privately owned and operated spacecraft to the

---

6. "Space Frontier Foundation," Space Frontier Foundation, accessed July 13, 2023, https://spacefrontier.org/.
7. Kelli Mars, "30 Years Ago: Daniel Goldin Sworn in as NASA's Ninth Administrator," Text, NASA History, March 31, 2022, http://www.nasa.gov/feature/30-years-ago-daniel-goldin-sworn-in-as-nasa-s-ninth-administrator.
8. "Space Tourism Society | Welcome," Space Tourism Society, accessed December 18, 2022, https://spacetourismsociety.org/.

edge of space, twice within two weeks.[9] The Ansari X PRIZE was able to successfully prove that contests do work, and inspired NASA and other groups to create competitions to give space entrepreneurs a chance to win funding for their projects. This is one of many examples of being at the right place at the right time to witness space history. The second project, the Yuri's Night World Space Party, is a global celebration about space travel.[10] I helped organize the first two San Francisco parties. I have to give a shout out to Loretta and George Whitesides for finding a way to mix space science education, creative arts, and music into a wonderful event that has spread worldwide. There even have been a few Yuri's Night parties onboard the International Space Station!

Besides my space activism addiction, I also developed a lot of hobbies. During the early 2000's, the San Francisco Bay Area had some amazing overlaps between engineering and the arts: robotics, software, industrial/kinetic arts, huge sculptures with fire effects. My friends David and Simone Calkins organized and ran ROBOGAMES, an international showcase of hobby robotics, which included many builders from the BATTLEBOTS TV Show.[11] It was truly a Robot Olympics: there was fighting, sumo, walking, soccer, fire-fighting, line following, hockey, art, and drink-making robots! Yes, I said drink-making robots, and I helped build three of them. These projects came in handy in learning about fluid flow, capillary action, and surface tension. You will learn more about this in an upcoming chapter.

Over the years, I got involved in the underground arts scene in the San Francisco Bay Area. There were some seriously awesome parties, many of them in costume! I stumbled upon Steampunk, a science fiction sub-culture focused upon re-envisioning the 19th Century Industrial Age in a more positive light, with inspiration from contemporary and vintage science fiction authors such as Jules Verne and H.G. Wells. Here is where I learned to build kinetic art vehicles and discovered the world of inventors and Maker Faires.[12] In the San Francisco Bay Area, many of these people and projects coalesced around an event called Burning Man.

I have learned about pushing the limits of my personal comfort zone by living in Black Rock City, a temporary community that is the core of the annual Burning Man event located in the Black Rock Desert in Nevada.[13] This art event is a grand experiment in both survival and fun. As with camping, there are no stores, no water, no supplies. Only what you bring with you. The dust storms make it hard to see, much less breathe. Daytime temperatures can reach up to 49°C (120°F). The "Playa," which is the Spanish word for "beach," is a lake bed in the Black Rock Desert that dries up in the summer. It is a remote, inhospitable place.

---

9. "XPRIZE Foundation Ansari Prize | XPRIZE Foundation," XPRIZE, accessed July 13, 2023, https://www.xprize.org/prizes/ansari.
10. "Yuri's Night—The World Space Party," Yuri's Night, accessed July 13, 2023, https://yurisnight.net/.
11. "RoboGames! (Formerly ROBOlympics)," accessed July 13, 2023, http://robogames.net/index.php.
12. "Maker Faire Bay Area," Maker Faire, accessed July 13, 2023, https://makerfaire.com/bay-area/.
13. "Burning Man—Welcome Home," Burning Man Project, accessed November 7, 2022, https://burningman.org.

Yet, somehow, nearly 80,000 people show up every year and convert the Playa into a showcase of massive art projects and experiments in alternative living. The word "Transformation" is brought up a lot at the event. That is because if you can somehow let go of your fears and old habits, you can make new, possibly better, habits for living your life. When you live in an extreme environment, you need to decide what to let go. After a while, you learn to improvise new habits which become new creature comforts or are similar to old ones. Imagination and creativity are very powerful things when you are faced with such circumstances.

The International Space Station is also a grand experiment in alternative living. Like Burning Man or a camping trip, there are many limitations to what you can do and enjoy out there. Unlike Burning Man, it is a government-run research laboratory, where science experiments take priority over comfort. Fortunately, commercial enterprises are coming in to bridge this gap. This is where you come in as a designer: how can you create a Space Lifestyle that is safe and more enjoyable for humans?

About the Structure of this Book

The first chapter will introduce you to the homestead analogy for off-world living: all the things that could potentially kill you in this strange new environment, and how you can re-envision them as design constraints for architecting a safer and more comfortable space habitat. With the fundamentals in place, I will then discuss in subsequent chapters key comforts: food and drink, bathrooms, toilets, showers, and finally the "kitchen sink" of comforts that need to be more thoroughly developed to help people feel more human in the Final Frontier.

Chapter 1: Dangers, Opportunities, Analogies introduces you to the challenges of life in space. You know it is dangerous out there and survival is the first priority. This chapter reviews the dangers that must be addressed and provides a context for how these dangers impact the topics in subsequent chapters. When dangers are viewed as design constraints, then opportunities emerge.

Chapter 2: Food and Drink discusses the challenges of creating and consuming cuisine and introduces my personal stories about "blob management." In addition to handwashing and kitchen appliances, topics include issues with hot and cold beverages, baked goods, hot meals, fresh foods, kitchen appliances, and bacon (a personal favorite).

Chapter 3: When Nature Calls covers the (number 1 and number 2) questions everyone asks astronauts. In a nutshell, space toilets suck—literally and metaphorically. From a design and usability standpoint, major improvements are needed to improve the quality of the toilet experience.

Chapter 4: Showers focuses on the evolution of the space shower. Designing a practical hot shower in space is a luxury that remains a challenge. This chapter provides a deep dive into its history and some ideas about possible solutions.

Chapter 5: Other Creature Comforts gathers together domestic topics that have limited research, but are extremely important to human comfort and enjoyment. Topics include keeping your space station clean, exercise and play, doing laundry, and the art of sleeping.

Conclusion wraps up what we learned with a challenge for you to adapt some of the efficiencies of off-world living here on Earth.

I hope this book whets your appetite to learn more about adapting domesticity to an extreme environment like space. Time to put on your designer hat, and explore the strangeness of off-world living!

Chapter 1:

Dangers, Opportunities, Analogies

Space travel is hard. Currently, off-world living is like an extreme camping trip. There are no gas stations, no hotels, no restaurants. You have to bring water, food, clothes, equipment, and shelter. Space is the harshest environment known to humanity. If the hard vacuum doesn't kill you, the radiation will. Or the occasional micrometeorite traveling a thousand times the speed of a bullet. NASA astronauts spend years training for their missions because of the dangers involved.

The time is long overdue to seriously improve the infrastructure to improve the safety for off-world living. In the 1970's and onward, after the success of NASA's Apollo missions to the Moon, both NASA and its Soviet/Russian counterpart, Roscosmos, took tantalizing steps into long term space habitation. Space Stations with names like Skylab, Salyut, MIR, and the International Space Station all became outposts in the wilderness. They were research laboratories: testing grounds for astronauts and cosmonauts to discover solutions to the challenges of microgravity, radiation, and other issues that would affect the human body. Astronauts spent many months in orbit around Mother Earth, protected by its magnetic field, and learned to live like space aliens. China has entered the space station adventure with its "Tiangong" space station.[14] They are also doing extensive science research similar to the USA, Europe, Japan and Russia, but are working independently. I believe it is finally time to take all of that space station experience and troubleshooting to design a better, more sustainable off-world lifestyle.

One of the benefits of the space programs is that many of the inventions and research eventually become new products and systems that benefit society. Imagine if we could reduce pollution dramatically by processing most waste products from the home, at the home? What if every home could be energy self-sufficient: by replacing fossil fuels with a safe micro power source such as a tiny nuclear generator? What if we can reinvent the ways we live so that human activity can be less damaging to the environment?

14. Andrew Jones and Daisy Dobrijevic, "China's Space Station, Tiangong: A Complete Guide," Space.com, August 24, 2021, https://www.space.com/tiangong-space-station.

Then there are folks talking about a "Plan B" for the human race: settling on the Moon and Mars as a backup for the human race if we damage the Earth too much. The plotline of the PIXAR movie "WALL-E" follows this idea too closely to the truth: a human society of mass consumption and mass distraction chooses to migrate into space and leave garbage-filled Earth, rather than take responsibility for the mess they have made.[15] I am not a fan of this vision. Why can't we settle on other planets AND clean up our home at the same time?

The International Space Station (ISS) is a good research lab for testing off-world living on a small scale, but it is time for the next step. Several private companies have proposed space stations to expand the human footprint in Earth's orbit, such as Above: Space Development Corporation (formerly Orbital Assembly), AXIOM Space, Blue Origin, GRAVITICS, Nanoracks, Northrop Grumman, and VAST Space. What will they put in them? How are they improving the human element? I refer to this throughout this book as the Space Lifestyle. How will these companies address issues such as food and drink, clothing and laundry, and showers and toilets? Have they considered or accommodated psychological needs, such as play and sleep? It would be nice to have a Star Trek level of convenience with replicators and holo-decks, but they have not yet been invented. The concepts proposed in this book are stepping stones to those conveniences.

Above: Space Voyager Space Station

Axiom Space Station

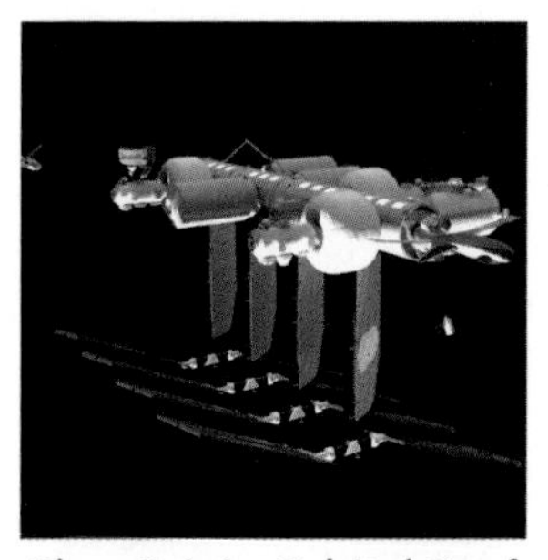

Blue Origin Orbital Reef

GRAVITICS StarMax

ThinkOrbital ThinkPlatform-4

VAST HAVEN-1

Voyager Starlab

Mixed-use private space stations proposed as of late 2023.
Note: This list is changing constantly.

15. "WALL-E," Pixar Animation Studios, accessed November 22, 2022, https://www.pixar.com/feature-films/walle.

Crossing fingers, I hope that at least one, if not several, commercial space stations are built and flying within 10 or 20 years. The technology of comfort and sustainability is here! We just need the right combination of design, engineering, business, marketing, and funding to make it happen.

As citizen astronauts continue to take short adventures to space, the demand for space access will grow. **Future space travelers need an improved quality of life, reduced exposure from the inherent dangers, and a simplified process of living off-world.** Support crews will need to live on board these next-generation space stations for months, which means the habitats will need to be much larger, store more supplies, have large food-growing areas, include a larger life support/ recycling system, and introduce entertainment spaces to relieve boredom and stress. In the process of building these larger habitats, we will need to develop new ways for sustainable living with less waste. This knowledge can be brought back to Earth to replace outdated and wasteful infrastructure with a more sensible system for living with nature, AND be used as a foundation for building larger space cities and homesteads on the Moon and the planets.

Spacesteading: Building a Homestead in Space

The homestead analogy is used in this book to describe the Space Lifestyle: a new way for long term, off-world living. The emphasis is on the creature comforts that make a house a good home and how that might apply in an extraterrestrial environment: quality food and drink, hot showers, entertainment, and more. **How do we make off-world living more enjoyable, and at the same time more sustainable?** We can pull the pieces of the puzzle together by looking at current research from NASA and private industry by reviewing "off-the-grid" living techniques on Earth and studying examples from history. From there, we can propose design concepts that make life in space more comfortable and easier to maintain.

What is a Home?

For most people, a home is a place with walls, windows, doors, and a roof. It is a place to put your stuff. It is a place to sleep, eat, drink, and go to the bathroom. It is where you live your life. For some, it is one of a series of houses bunched together in subdivisions, near roads, stores, and places to eat. For others, it is a small apartment in a high rise, along with dozens of similar places.

Imagine you lived on a farm, miles from civilization. Where does the power come from? Where do you get water? When you flush the toilet or take a shower, where does that wastewater go? Where does the food come from? How can you live comfortably? How much space do you need to live?

In space you have a similar situation to farm life or the homestead, compounded by all the dangers mentioned in the section below. Onboard the International Space Station (ISS), astronauts are tending several experimental farms in progress, but not enough food can be grown to sustain the crew independently without support from Earth. Before we discuss the details of homesteading in space, let's understand what we are dealing with.

How Not to Die in Space: Know Your Environment

For newcomers to the space industry, get ready for a wild ride. The following concepts we're going to talk about are unique to the space environment. Some of these issues may be hard to grasp. I will walk you through each one and give you examples, and discuss possible design options to deal with the issue. Below I will describe each issue, list the dangers, and then discuss current and potential design solutions.

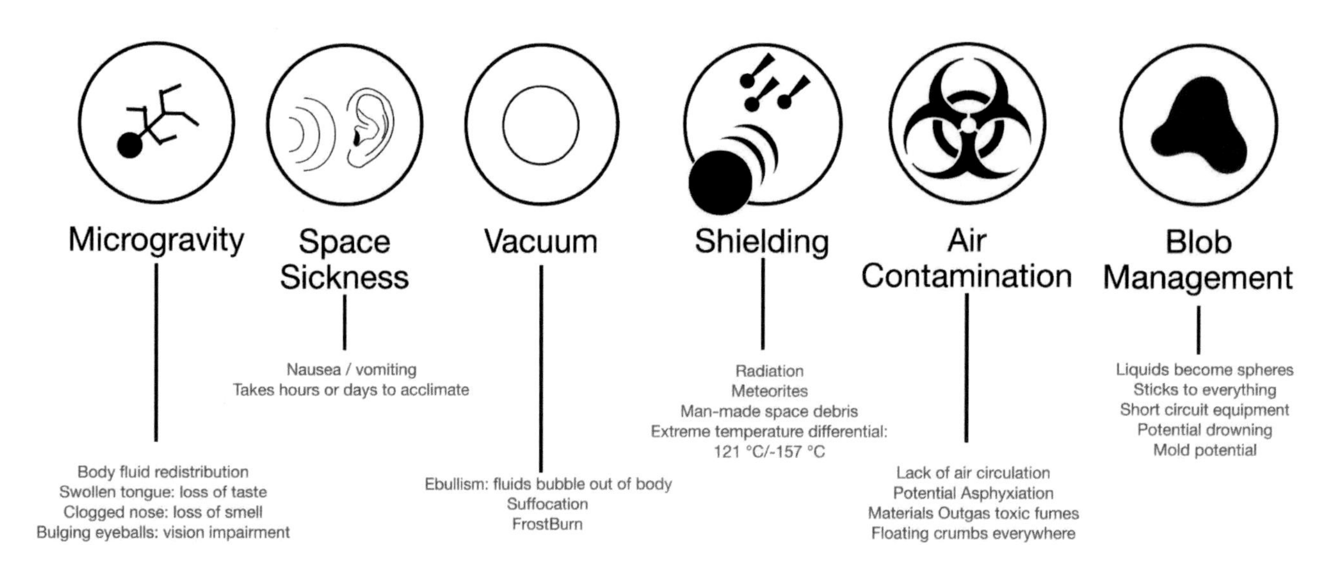

Living in space has many challenges. Images by the author.

Microgravity

This is everybody's favorite topic. It appears like floating in water, but it's more complicated than that. A better analogy is riding on a roller coaster. After you reach the top of the roller coaster hill, you start going down it. That is the point where the blood rushes to your head and upper body. You begin to float, yet you feel that sensation of falling. You just never stop. You don't reach a "destination," like the ground. Another way to think about it is hilariously described in a science fiction book by Douglas Adams called "Life, the Universe and Everything."[16]

> *"The Guide says there is an art to flying, or rather a knack. The knack lies in learning how to throw yourself at the ground and miss."*

Officially called "microgravity," this sensation is created by falling in such a way that gravity's effects are greatly diminished. Gravity is still trying to pull you into the Earth, but the spaceship or space station you are traveling in is moving so fast it misses it![17] Thus: continuous falling.

---

16. Douglas Adams, *Life, the Universe and Everything* (London: Pan Books Ltd., 1982).

17. Sandra May, "What Is Microgravity?," NASA, *What Is Microgravity?* (blog), February 15, 2012, https://www.nasa.gov/audience/forstudents/5-8/features/nasa-knows/what-is-microgravity-58.html.

Dangers: A few days of weightlessness can be fun, but after a while the physical effects to the body become apparent.

- Without gravity, your blood redistributes evenly across your body. Your face looks like a balloon! It is a similar effect to hanging upside down in a jungle gym, except it is ongoing no matter what your body orientation is.
- Here are a few of the issues you will start to experience the longer you are in a microgravity environment:
    - You lose your sense of taste because of a swollen tongue.
    - Your eyes start to lose focus because of the blood pressure.[18]
    - Your bones start to lose calcium because your body no longer needs to hold up your weight while you stand or sit.

Did I mention that legs are irrelevant in space? People with one or no legs can actually maneuver around more easily in a spacecraft or space station. There is even a group called AstroAccess promoting this concept, and helping give people with disabilities a chance to fly on ZERO G parabolic aircraft.[19]

Design Solutions: Currently astronauts and cosmonauts do an astonishing TWO HOURS of exercise every day to reduce the impact of bone loss. It is not enough, especially for 6 or more months in orbit. Three possible combinations of solutions could help with this:

- **Diet:** Develop a calcium-enriched diet to help the space traveler eat better and increase the mineral in the body.
- **Exercise:** Build a dedicated gymnasium with a much larger play area to create new types of sports and exercises.
- **Gravity:** Build a variable-gravity space station. Several companies, including ABOVE: Space Development Corporation, Gravitics, gravityLab, and VAST Space, have plans for building space stations or satellites that can increase or decrease its spin to generate up to half Earth Gravity (0.55G).[20] Space stations like this will be a revolution in space habitat design, if they can be built and proven to work. Off-world life would be so much easier for humans if their bodies didn't deteriorate so quickly.

---

18. Peta Bradbury et al., "Modeling the Impact of Microgravity at the Cellular Level: Implications for Human Disease," *Frontiers in Cell and Developmental Biology* 8 (February 21, 2020), https://doi.org/10.3389/fcell.2020.00096.
19. "AstroAccess Successfully Completes 1st Weightless Research Flight with Int'l Disabled Crew—SatNews," accessed October 29, 2023, https://news.satnews.com/2022/12/16/astroaccess-successfully-completes-1st-weightless-research-flight-with-intl-disabled-crew/.
20. "ABOVE: Space Development Corporation," ABOVE: Space Development Corporation, accessed October 29, 2023, https://abovespace.com/.

Space Sickness

Similar to air sickness or seasickness, the human body can be very sensitive to the new sensations of being off-world. The vestibular system in the inner ear helps your body with balance and orientation. Sudden changes in motion or gravity can cause you to get sick, and it is unpredictable. As mentioned in the above description about microgravity, your body feels like it's on a continuous roller coaster ride. The sensation of free fall is continuous in space. As the vestibular system in your inner ear helps your body with balance and orientation the extra fluids flowing to your head the pressure upon your ears and cochlea could cause intense nausea. After a few hours most people get used to it, but some folks struggle for quite a while.[21]

Speaking of vestibular systems, did you know that deaf people don't get motion sickness? In the very early days of the Space Race, NASA and the US Navy did studies of the vestibular systems of 11 deaf students from Gallaudet University. They were subjected to parabolic airplane flights, slow rotation rooms, and even a trip on an ocean ferry. The researchers got more sick than the students! Apparently, the genetic or accidental damage to the deaf students' hearing had also affected their vestibular system, which detects motion. The deaf students could endure the intense motion changes without any ill effects.[22] Deaf astronauts would have a unique advantage by handling orientation changes in space. Maybe a device or surgical procedure could be developed to help the rest of us be less sick on long term space flights.

Dangers: Officially called Space Adaptation Syndrome (SAS) by NASA, it is mostly unavoidable, especially during the first few days. For some astronauts, they get mildly nauseous. Others tend to vomit up until their body gets used to that roller coaster floating sensation. A legendary space sickness incident occurred during the STS-51D Space Shuttle mission in 1985, where Senator Jake Garn, who was officially flying as a Congressional Observer, vomited so much that the astronauts used his condition as an unofficial unit of measurement for how sick a person could be. For example, most people get 1/10th Garn sick during a spaceflight.[23]

Design Solutions: This is a hard one. Even veteran fighter jet pilots who became astronauts can suffer from SAS. Here are some suggestions from NASA:

- **Focus:** A person can focus on one spot, like a book, or the wall, and wait it out.
- **Medication:** Anti-nausea medication might work, except in cases where it does not.
- **Wait it out:** Unfortunately, the human body takes time to adjust to space travel.

---

21. Patrick L. Barry and Tony Phillips, "Mixed Up in Space," NASA, Science @NASA Blog, August 7, 2001, https://web.archive.org/web/20090513111327/http://science.nasa.gov/headlines/y2001/ast07aug_1.htm.
22. Andres Almeida, "How 11 Deaf Men Helped Shape NASA's Human Spaceflight Program," NASA, May 4, 2017, http://www.nasa.gov/feature/how-11-deaf-men-helped-shape-nasas-human-spaceflight-program.
23. Carol Butler, "Robert E. Stevenson Oral History," NASA Johnson Space Center, May 13, 1999, https://historycollection.jsc.nasa.gov/JSCHistoryPortal/history/oral_histories/StevensonRE/StevensonRE_5-13-99.htm.

Vacuum

The obvious fact is there is no air up there in space. The higher up you go in Earth's atmosphere, the less air there is. The atmosphere gets thinner and thinner, and when you reach the height of the ISS (250miles/100km), the air pressure is nearly 0 Pascals (or 0 PSI). The lack of air resistance at higher altitudes makes it easier to travel around, but obviously does not allow you to breathe. When you start to travel between the planets, you will encounter a near-perfect vacuum, with only a handful of molecules per square meter.

Dangers: Yes, it is possible to survive in space without a space suit, for about 10 to 15 seconds that is.[24] Here is what happens your body in a vacuum (trigger warning: skip the next part if you are squeamish):

- **Ebullism:** Ebullism is when bodily fluids form gas bubbles because of reduced pressure.[25] This means that not only would the nitrogen cause your skin to bubble up, but all of your bodily fluids will too. This consists of tears, saliva, and any liquid found elsewhere in the body. The worst thing to do is hold your breath. Due to the lack of pressure in space, the oxygen located inside the lungs would expand causing the lungs to rupture.
- **Sunburn and Frostbite**: Without the protection of the Earth's atmosphere, the UV rays, X-rays, and gamma rays would not only burn your already bubbling skin, but also damage your DNA (more on this later). This means that if survival was miraculously achieved you would likely end up with not only cancer but a horrendous sunburn. At the same time, your body heat will dissipate and your body slowly starts to freeze: somewhere close to -157 °C (-250.6 ºF).
- **Frozen forever:** Without oxygen, the bacteria in your body couldn't start the decomposition process. You would basically be a frozen mummy for all time.

Design Solutions: Don't open the airlock. Duh! Seriously.

- **Airlocks:** NASA has some pretty reliable airlock systems. The latest one, built by the company Nanoracks, is the Bishop Airlock Module. It is designed for larger payloads to enter and exit the ISS. In fact, the entire airlock can disengage with the space station! This allows it to double as a launcher of payloads such as cubesats (super small satellites, and for launching garbage back to Earth for burning up.[26] For future space stations, where private astronauts

---

24. Abigail Edwards, "Can Humans Survive in Space Without a Space Suit?," *Penn State University, SiOWfa16: Science in Our World: Certainty and Controversy* (blog), September 11, 2016, https://sites.psu.edu/siowfa16/2016/09/11/can-humans-survive-in-space-without-a-spacesuit/.
25. Daniel H. Murray et al., "Pathophysiology, Prevention, and Treatment of Ebullism," *Aviation, Space, and Environmental Medicine* 84, no. 2 (February 2013): 89–96, https://doi.org/10.3357/asem.3468.2013.
26. Margo Pierce, "A New Doorway to Space," Text, NASA, November 13, 2020, http://www.nasa.gov/directorates/spacetech/spinoff/New_Doorway_to_Space.

with less extensive training are onboard, there needs to be better safety mechanisms to avoid accidental decompression.

- **Spacesuits:** Besides NASA's venerable spacesuits (which have not changed much since Apollo and the Space Shuttle programs), SpaceX has designed some sleek, new spacesuits for their Dragon spacecraft.[27] Also, there have been some amazing alternative space suit designs by Dava Newman's team at MIT's Media Lab.[28] The MIT spacesuit is thinner, easier to put on, and full of embedded sensors. Private company AXIOM Space has teamed up with fashion design company PRADA to develop a new spacesuit based upon NASA designs, but with a modern design style that hopefully will replace NASA's clunky space suits.[29]
- **Single Person Spacecraft (SPS):** Using a concept that goes way back to the 1950's and space pioneer Werner von Braun, the company Genesis Engineering Solutions is developing a Single Person Spacecraft as an alternative for bulky spacesuits.[30] Typically, astronauts have to spend hours of preparation, including pre-breathing pure oxygen in an airlock, before they can even get into their spacesuits. With the SPS, you can simply float into the spacecraft chamber, close the airlock, and jettison from the space station and start to work. Robot arms on the outside of the craft are directly controlled by the pilot. The advantages are dramatic: in emergencies, you can quickly get outside and do repairs without waiting. The gloved hands of a space-suited astronaut get tired very easily, so robot hands make more sense for handling objects and doing detailed tasks. Also, private astronauts do not need to go through the extensive spacesuit training that NASA astronauts go through to take a "spacewalk." [31] All they have to do is hop in an SPS and fly around.

---

27. Elizabeth Howell, "New SpaceX Spacesuits Get Five-Star Rating from NASA Astronauts," Space.com, June 10, 2020, https://www.space.com/spacex-spacesuits-five-star-astronaut-review.html.
28. Sarah Beckmann, "Dava Newman Presents 3D Knit BioSuit™ at 2022 MARS Conference," MIT Media Lab, March 30, 2022, https://www.media.mit.edu/posts/dava-newman-presents-3d-knit-biosuit-at-mars-conference/.
29. "Axiom Space, Prada Join Forces on Tech, Design for NASA's Next-Gen Lunar Spacesuits," Axiom Space, October 4, 2023, https://www.axiomspace.com/news/prada-axiom-suit.
30. Brand N. Griffin et al., "Single-Person Spacecraft Provides Commercially Viable EVA Including Tourist Excursions for Orbital Reef" (ASCEND 2022, Las Vegas, Nevada: American Institute of Aeronautics and Astronautics, 2022), 8 pages, https://doi.org/10.2514/6.2022-4209.
31. Brand N. Griffin et al., "The Wait-Less EVA Solution: Single-Person Spacecraft," ASCEND (ASCEND 2020, Virtual Event: American Institute of Aeronautics and Astronautics, 2020), https://doi.org/10.2514/6.2020-4170.

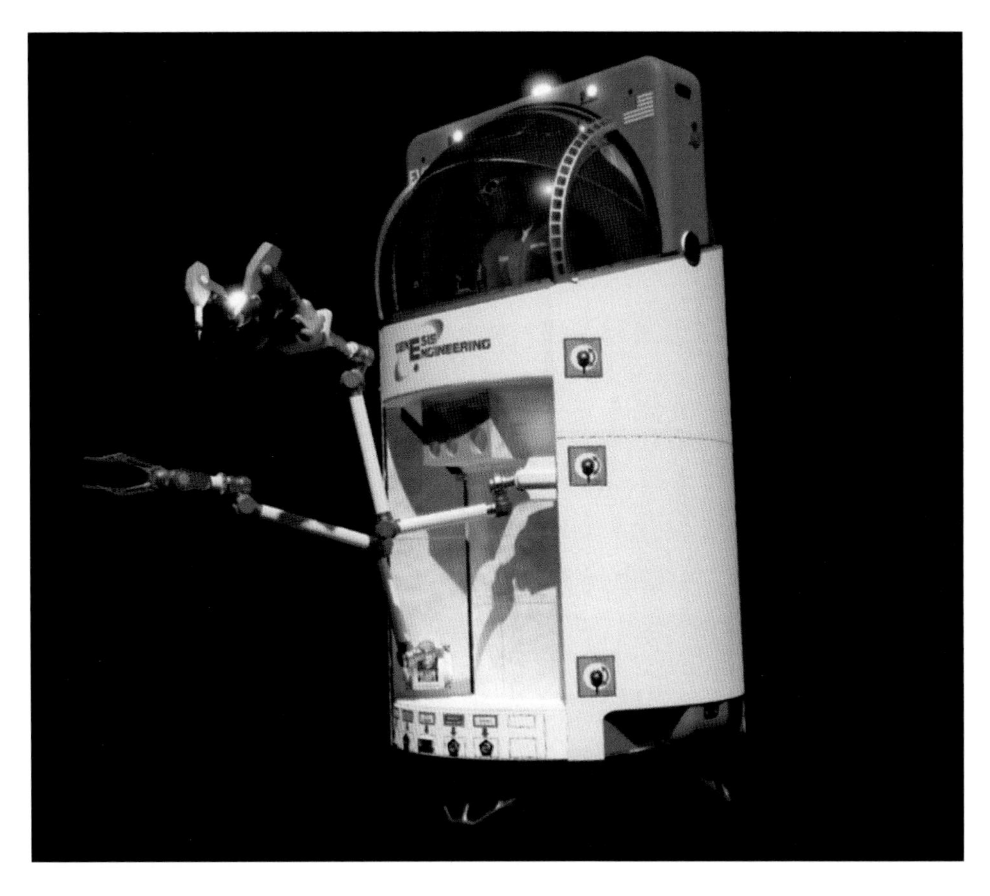

The Single Person Spacecraft, by Genesis Engineering, could revolutionize Extravehicular Activities (EVAs) with a mini-spaceship for doing maintenance outside of space stations. Source: Genesis Engineering Solutions.

Radiation, Rocks and Debris

We are blessed to have a strong atmosphere and magnetic field around our planet.[32] These two phenomena absorb or reflect most of the space rocks and radiation coming from the sun and elsewhere from reaching the surface of the Earth. They are our natural space shield.

The farther away from the surface of the Earth you go, the more exposed you become to all kinds of dangers. Several NASA scientists propose the only way to safely get to Mars is by going really fast to reduce the radiation damage to the crew. While traveling fast is ideal, you cannot run away from radiation and its long term harm to people. If you think you will be safe when you get to Mars, you are in for a surprise: Mars' magnetic field was smashed millions of years ago, so you would need to bring our own along.[33]

32. Alan Buis, "Earth's Magnetosphere: Protecting Our Planet from Harmful Space Energy," *NASA: Global Climate Change: Vital Signs of the Planet (blog)*, August 3, 2021, https://climate.nasa.gov/news/3105/earths-magnetosphere-protecting-our-planet-from-harmful-space-energy.
33. "Did Mars's Magnetic Field Die With a Whimper or a Bang?," *Science* (blog), April 30, 2009, https://www.science.org/content/article/did-marss-magnetic-field-die-whimper-or-bang.

Dangers: Without our home planet's natural shielding, all sorts of nasty things can come at you:

- **Radiation:** Some of the varieties of radiation you can find in space include gamma rays, cosmic rays, UV rays of all types, and more. The most common effect of this is the damage to your DNA and the central nervous system as a ray blasts apart molecular chains in your body. Sometimes the body can repair the damage, but other times it repairs the damage incorrectly, causing a mutation. This mutation could lead to cancer.[34] Like getting x-rayed in a hospital, your body can recover from short doses. But your body is not designed to handle the constant radiation of that same x-ray machine 24 hours a day, seven days a week.
- **Meteoroids:** Meteoroids are everywhere in space![35] Predicting when one will strike a spacecraft is extremely difficult. Rocks and debris of any size are dangerous to space station hulls and space suits. Meteoroids could be traveling at immense speeds. Some have been clocked at speeds of 71 kilometers (44 miles) per second through the Earth's atmosphere...and that is with atmospheric drag slowing it down![36] Like a bullet through tissue paper, a pebble could rip through the side of a spacecraft very easily.
- **Man Made Space Debris:** There are also a large number of dead satellites and rocket bits floating around Earth's orbit. More than 27,000 pieces of orbital debris, or "space junk," are tracked by the Department of Defense's global Space Surveillance Network (SSN) sensors.[37] [38] There have been several close calls in recent years when the ISS needed to make an emergency boost to a different orbit when space debris got too close.[39]
- **Temperature Differential:** The temperature on the side of the International Space Station (ISS) facing the Sun can reach 121 °C (250 ºF). The side facing away from the Sun can go as low as -157 °C (-250.6 ºF). This is a difference of almost 300°C (572 ºF)![40]

So how do you shield yourself from all these nasty things?

---

34. Kelli Mars, Amy Blanchett, and Laurie Abadie, "Space Radiation Is Risky Business for the Human Body," Text, *NASA* (blog), September 18, 2017, http://www.nasa.gov/feature/space-radiation-is-risky-business-for-the-human-body.
35. "10 Things: What's That Space Rock?," *NASA Science*, July 21, 2022, https://science.nasa.gov/solar-system/10-things-whats-that-space-rock/.
36. Jeannie Evers, "Meteor," *National Geographic Society*, October 19, 2023, https://education.nationalgeographic.org/resource/meteor.
37. Mark Garcia, "Space Debris and Human Spacecraft," Text, *NASA*, April 13, 2015, http://www.nasa.gov/mission_pages/station/news/orbital_debris.html.
38. Edward P. Chatters et al., "Space Surveillance Network," *AU-18 Space Primer (Air University Press, 2009)*, https://www.jstor.org/stable/resrep13939.26.
39. ScienceAlert Staff, "A Chunk of Satellite Almost Hit The ISS, Requiring an 'Urgent Change of Orbit,'" *ScienceAlert*, November 15, 2021, https://www.sciencealert.com/a-chunk-of-chinese-satellite-almost-hit-the-international-space-station.
40. "Temperature on Earth and on the ISS," *Let's Talk Science*, September 23, 2019, https://letstalkscience.ca/educational-resources/backgrounders/temperature-on-earth-and-on-iss.

Design Solutions: The most ideal solution would be to build a thick-walled hull to protect against all these dangers. Unfortunately, at the current time the cost of launching such a heavy mass is impossible. When SpaceX finally starts flying the Starship spacecraft, the cost of flying larger, heavier structures should drop dramatically. Below is a list of design options, from those currently being used, to longer term concepts.

- **Reflective blankets:** The ISS uses layers of insulating fabric to moderate the temperatures. Most of the ISS is coated with blankets of Multi-Layer Insulation (MLI), which are made up of layers of Mylar and Kapton, the same shiny stuff used in emergency blankets here on Earth (which is a spinoff tech from NASA's Apollo program). Not only does the material reflect back the excess heat, it also helps protect from solar radiation.
- **Radiators:** Also on the ISS, they use Heat Rejection Subsystem (HRS) radiators. These radiate (transfer) excess heat built up inside the ISS into space.
- **Whipple Shields:** Armor shielding spaced a certain distance apart from each other are called spaced armor, or Whipple Shields. Spaced Armor is used in many cargo ships and some military tanks.[41] Each layer reduces the damage of a high speed object such as bullets, micrometeoroids or space junk. (Note: They are called micrometeorites when the particle survives going through Earth's atmosphere and reaches Earth's surface). Each layer is not expected to stop the particle, but to break it up into smaller pieces, dividing up the energy into smaller and smaller chunks. When the tiny particles are spread more thinly across a larger wall area, there is a better chance for each successive wall to withstand the impact. "Stuffed" Whipple Shields include Kevlar or Nextel aluminum oxide fiber in between the walls, improving damage reduction.[42]
- **Water Walls:** NASA Ames Research Center developed the "Water Wall" concept to not only supply and filter water for the astronauts, but the placement of the water and Polyethylene bags around the inner hull of the space station offer dramatic radiation protection.[43] Water Walls are discussed in more detail later in this chapter.
- **Asteroids:** For longer term space travel to Mars and other places far from Earth, you are going to need thicker insulation from cosmic radiation and random particle impacts. Several engineers proposed capturing an asteroid and building a space station inside it.[44] Hiding in

---

41. A Hurlich, "Spaced Armor" (*Second Tank Conference, Ballistic Research Laboratories*, Aberdeen Proving Ground, Aberdeen, Maryland: Watertown Arsenal Labs, 1950), 24, https://apps.dtic.mil/sti/citations/ADA954865.
42. A. Pai et al., "Advances in the Whipple Shield Design and Development:," *Journal of Dynamic Behavior of Materials* 8, no. 1 (March 1, 2022): 20–38, https://doi.org/10.1007/s40870-021-00314-7.
43. Michael Flynn et al., "Membrane Based Habitat Wall Architectures for Life Support and Evolving Structures" (40th International Conference on Environmental Systems, Barcelona, Spain: American Institute of Aeronautics and Astronautics, 2012), https://doi.org/10.2514/6.2010-6073.
44. Thomas I. Maindl, Roman Miksch, and Birgit Loibnegger, "Stability of a Rotating Asteroid Housing a Space Station" (arXiv, December 26, 2018), https://doi.org/10.48550/arXiv.1812.10436.

a big rock is natural insulation from radiation, impacts, and temperature extremes. There have been concepts in science fiction about building a spaceship inside a small asteroid. The biggest challenge is logistics. Very interesting concept but still far from being possible.

Air Contamination

Like a being in a stuffy closet, air stands still in microgravity. Without gravity there are no convection currents to naturally circulate the air. On top of that, humans are messy. We make messes every day. At least on Earth, if you drop something, it falls to the floor and you can clean it up. Cleaning your room in space is a whole new challenge. In a microgravity space station, crumbs, dirt, and even dust will just float around you. Many foods we enjoy on Earth, like cookies or bread, make lots of floating crumbs. NASA has banned bread, cookies, and cakes due to their crumbly nature.

Dangers: There are three possible dangers to humans without proper air flow and filtration:

- **Asphyxiation:** It is quite possible that in your sleeping quarters you could consume all of the oxygen and be surrounded by carbon dioxide. On Earth, normal air flow keeps oxygen and CO2 flowing freely. Astronauts sleep with fans in their quarters just to keep the airflow going properly.
- **Outgassing:** That new car smell. The taste water takes on after being stored in a plastic container for a time. The smell of a freshly painted room. These recognizable smells and tastes are all due to outgassing. Outgassing or off gassing is a normal process for many manufactured products.[45] On Earth, you can simply open the window or turn on a fan to make the smell go away. In a cramped space station with no convection currents to move the air, those smells stay in one place, and in high concentrations could be poisonous.
- **Crumbs Everywhere:** Finally, without a good air circulation and filtration system to capture airborne debris, the spacecraft interior will become clogged and crusty and filled with little bits of just *stuff*, and this will become unlivable over time.[46]Also, if something breaks or shatters, the pieces are extremely difficult to clean up. There is no "sweeping it under the rug." You must clean up all messes or people could die.

Humor time: ever heard of a "fart bomb?" In a spaceborne version of the joke "He who smelt it, dealt it," it is possible to prank someone by farting in one location, and someone else will float right through the invisible cloud of ick. All humor aside, without airflow, that stinky cloud of methane could stay in that one place in the space station forever![47] Thanks to my experience playing with children for that one.

45. Jennifer Mathias, "What Is Outgassing Testing?," *Innovatech Labs* (blog), January 26, 2016, https://www.innovatechlabs.com/newsroom/882/outgassing-testing/.
46. Sandra May, "Eating in Space," Text, NASA, June 8, 2015, http://www.nasa.gov/audience/foreducators/stem-on-station/ditl_eating.
47. Ryan F. Mandelbaum and Rae Paoletta, "We Chatted With an Astronaut About Showering, Farting, and Boning in Space," Gizmodo, April 21, 2017, https://gizmodo.com/we-chatted-with-an-astronaut-about-showering-farting-1794538749.

Design Solutions: Ideally, you need continuous airflow, such as from fans or convection.

- **Materials:** NASA has very stringent materials outgassing requirements.[48] They learned very early on that materials outgassing in a sealed spacecraft is dangerous to the crew.
- **Air vents:** For microgravity space stations, every room (including closets) needs to have dedicated air intake and outtake vents, to keep the air fresh and safe from poisonous accumulations of any gas or particles.
- **Filters:** Besides air circulation, these vents need a filter system to capture random bits of debris, such as food particles, that tend to float randomly all over the place. Each of these air intakes need to have a filter, or at least a fine mesh screen on the interface. These filters need to be cleaned out on a regular basis, like is currently happening onboard the ISS.[49] Later in this chapter, I will talk about NASA's life support system that manages the air and water filtration and recycling.
- **Gravity:** If a variable gravity space station could be built, that would help generate some natural airflow due to gravity pulling and moving the air. It would also make the station much more comfortable for humans to live in.

Liquid Contamination

Water is the key to not only the basic creature comforts such as drinks, showers, or going to the bathroom, it is also the key to survival. Like the discussion about the lack of airflow above, there is also the lack of water flow. Water, oils, and other liquids do strange things without gravity.

On Earth, surface tension is the natural tendency of the molecules on the outer edges of a liquid to stay close to similar molecules. You can see the curved shape of a drop of water on glass, or any non-absorbent surface. Oils, soap bubbles, or most any liquid act in this manner. Similar molecules stick together better at the interface between liquids and gasses. This is called surface tension.[50] Liquids like to stick to themselves and the object they are in contact with. With the exception of soap bubbles, most liquids have too much mass to fight against gravity and tend to flatten.

Without gravity, liquids naturally turn into spheres or blobs. The cohesive nature of water, which you barely notice when you wash your hands on Earth, becomes much more obvious in space. There is an excellent video of astronaut Chris Hadfield trying to squeeze water from a washcloth.[51] The water

48. "Description | Outgassing," *NASA Goddard Space Flight Center*, accessed October 23, 2022, https://outgassing.nasa.gov/Description.
49. *Chris Hadfield—Nail Clipping in Space*, Canadian Space Agency (International Space Station, 2013), https://www.youtube.com/watch?v=xlCkLB3vAeU.
50. Jyrki Korpela, "What Is Surface Tension?," *Biolin Scientific* (blog), October 16, 2018, https://www.biolinscientific.com/blog/what-is-surface-tension.
51. *Wringing out Water on the ISS—for Science!* (International Space Station: Canadian Space Agency, 2013), https://www.youtube.com/watch?v=o8TssbmY-GM.

tenaciously stuck to the outside of the cloth and started to envelop his hands! The behavior of water in space will come up several times in this book, as blob management is critical for creature comforts, survival, and sustainability.

Surface Tension

Without gravity, water sticks to everything!

Examples of water blobs and surface tension of water in microgravity. Image sources: NASA.[52]

Dangers: It's dangerous when water doesn't behave like you expect it to, like it does on Earth.

- **Equipment Damage:** Sticky blobs can get into electronic equipment and cause a short circuit.
- **Drowning:** One potential danger of lack of water flow is in the case of someone taking a shower, or most especially a bath, in microgravity. Enough water could quite possibly envelop a person's face and cause them to suffocate. The water is hard to wipe off and could become fatal.

Design Solutions:

- **Pumps:** Peristaltic water pumps move liquids from place to place in a controlled manner.[53] These pumps are used in drink vending machines and scientific equipment like DNA analyzers.
- **Slippery Surfaces:** Hydrophobic materials, such as certain plastic and paint coatings, repel water and prevent clinging.[54] If used in sinks, showers, and certain gardening containers, you can manage water more easily.

---

52. Shelley Canright, "The Physics of Space Gardens," NASA Education (Brian Dunbar, June 14, 2003), https://www.nasa.gov/audience/forstudents/5-8/features/space_gardens_feature.html.
53. "How Does A Peristaltic Pump Work?," *TapFlow Pumps UK*, accessed November 7, 2023, https://www.tapflopumps.co.uk/blog/peristaltic-pump-guide/.
54. "Hydrophobic Substances: What Are They and What Are They Used For?," *Infinitia Research* (blog), May 26, 2021, https://www.infinitiaresearch.com/en/news/hydrophobic-substances-what-are-they-and-what-are-they-used-for/.

So, with a list of challenges like this, do you still want to go into space?

How to Survive in Space: Design Analogies

It's time to repackage these issues as design constraints. As a designer you see the world differently. You can see these things that would normally be considered life-threatening as a possible opportunity. By understanding the factors of your situation, you can make logical choices as to what materials should be used, what technologies could be developed, and what lifestyle adaptations need to be made. Remember, you have just left the Earth. Everything is different.

The homestead you are going to live in will not look like a typical farm, with livestock, hay bales, and rows and rows of corn stalks reaching for the sky. What you are going to live in will be much more compact, modular, and optimized for sustainability. Now we can talk about Space Architecture.

What is Architecture? It is more than just drawing lines on a piece of paper or a Computer Aided Design (CAD) software on a computer. It is about living and building a place that caters to the way you live. That means looking at the aesthetics of an environment, and finding ways to work with it. You can enhance the beauty of one thing and reduce the inconveniences of another thing. And of course, you can add some flair to it. A house does not just have to be a box shape. It could have curves or spires or other fanciful elements of your choosing.

What the heck is Space Architecture? Basically, it is architecture in an extraordinary environment with unusual design requirements. Since 2002, the Space Architecture Technical Committee (SATC) of the American Institute of Aeronautics and Astronautics (AIAA) has promoted educational programs in the field of Space Architecture.[55] I was part of that inaugural meeting in 2002, and have interacted with many space architects over the years. I recommend starting at the *Spacearchitect.org* website and learn more about this fast growing field.

For the rest of this section, I am going to present to you with design analogies, Earthbound experiments, and some space design concepts to help you understand what a homestead in space could look like.

Why use analogies? Because you, as a designer, will need to understand the complexities of off-world living. For example, an airlock chamber could be analogous to a "mudroom" or foyer in a house: sort of a transitional entrance from the harsh outside "weather" to the warm and comfortable indoors.

Let's start with Earthbound examples, and then move out to orbit.

55. SPACEARCHITECT.ORG, May 17, 2019, https://spacearchitect.org/.

A House Inside a Greenhouse

An architect in Belgium built what he calls the "KASECO" house...inside a greenhouse.[56] This experimental greenhouse contains enough plant material to sustain a family in a house inside of it. The advantage of this design is that the temperature is under control at all times inside the greenhouse. If it gets too warm, vents open on the roof. As part of sustainability, solar panels and batteries collect electricity from the sun. Rainwater is collected on the sides of the house by catch basins, which is circulated into filtration systems inside the house for drinking water. It is a highly efficient system. Even the wastewater from sinks, showers, and toilets are filtered through certain plants in the garden.[57]

KASECO greenhouse built in Belgium. Source: Screen shot from "Living in a Garden Oasis: The Delights and Surprises of Giant Greenhouse Living," Mindful Building and Living YouTube Channel, Jun 1, 2021. "Kaseco-concept" by Architect Koen Vandewalle. Tel: 0032 477 35 49 42, Email: Koen@kaseco.be.

56. Koen Vandewalle and Samia Wielfaert, "Kaseco | The Ultimate Greenhouse," Kaseco Plus, accessed October 26, 2022, https://www.kaseco.plus/en.

57. *This Big Greenhouse Built around a House Is the Home for a Family of 7* (Murissonstraat, 8930 Rekkem (Menen), Belgium, 2021), https://www.youtube.com/watch?v=atc6-JCVIOs.

Biosphere 2: A Space Homestead Prototype

Another example comes from a desert experiment from the 1990's called Biosphere 2.[58] Seen by the news media at the time as a weird cult project, it was actually a magnificent experiment in global ecology for off-world living.[59] Built just outside of Tucson, Arizona, USA, the place covers over 3 acres, sealed from the bottom by stainless steel and covered on the top by an airtight glass space frame structure. Inside the laboratory is a 850 square meter coral reef, a 450 square meter mangrove marsh, a 1,900 square meter Amazonian rainforest, a 1,300 square meter savannah grassland, a 1,400 square meter fog desert, and a 2,500 square meter tropical agriculture system for farming. The human habitat features living quarters, offices, and recreational spaces. The heating and cooling water is circulated through Biosphere 2 via independent piping systems. Electrical power was supplied from a natural gas energy center, located outside Biosphere 2, through airtight penetrations. The results were amazing:[60]

- The massive closed ecosystem was larger than anything NASA has built for life support experiments.
- The eight member crew stayed inside for two years, a record for any enclosed experiments at the time.
- The facilities' annual air-leak rate of less than 10 percent is the lowest leak rate of any such structure ever built.
- The tight seal enabled the detection of an oxygen decline, which has led to significant research about oxygen cycles.
- The crew produced approximately 80 percent of its food, setting another world record for food production in a closed system, despite two winters of record high rainfall and cloudy weather. The remaining approximately 20 percent was provided from seed stock and from crops grown and stored in Biosphere 2 prior to closure.
- The crops were produced without the use of toxic pesticides or chemical fertilizers.
- The ocean system was the largest artificial coral reef in the world and sustained over 800 coral colonies with 87 new baby colonies formed during the two years.

---

58. Mark Nelson, "The Challenge of Managing Water and Nutrient Cycles in a Mini-World—the Lessons from Biosphere 2," *Global Water Forum* (blog), August 6, 2020, https://globalwaterforum.org/2020/08/06/the-challenge-of-managing-water-and-nutrient-cycles-in-a-mini-world-the-lessons-from-biosphere-2/.
59. *Jane Poynter: Life in Biosphere 2*, 2009, https://www.youtube.com/watch?v=a7B39MLVelc.
60. Mark Nelson, "Projects—Biosphere 2—The Institute of Ecotechnics," *Institute of Ecotechnics* (blog), 2, accessed October 26, 2022, https://ecotechnics.edu/projects/biosphere-2/.

The layout of Biosphere 2. Source: Dr. Mark Nelson, ecological engineer for Biosphere 2.

Although the Biosphere 2 experiment goes way beyond the needs for the homestead in space that I am describing, it is definitely a role model for the next few generations of space stations that we should aspire to design and build.

Next, let's take a look at what NASA is doing to create a sustainable life support system onboard the International Space Station.

NASA's Life Support System

On Earth, many systems in nature process air and water. Currently onboard the International Space Station, these things are re-created artificially via mechanical and chemical means. NASA uses the Environmental Control and Life Support System (ECLSS) to emulate the Earth's natural life support system.[61]

61. Lee Mohon, "Environmental Control and Life Support System (ECLSS)," Text, *NASA*, September 11, 2017, http://www.nasa.gov/centers/marshall/history/eclss.html.

- It performs several functions:
- Provides oxygen for breathing
- Removes carbon dioxide from the cabin air
- Recovers and recycles oxygen from carbon dioxide to resupply the crew
- Filters particulates and microorganisms from the cabin air and maintains cabin pressure, temperature and humidity levels
- Removes volatile organic trace gasses such as ethanol, that are colorless, odorless and can build up over time
- Distributes cabin air between each room, or module, of the station
- Provides potable water for consumption, food preparation and hygiene
- Purifies recycled water from multiple sources back to potable water

NASA is constantly tinkering to improve the reliability of ECLSS so that one day it can support crews on the Moon and Mars. Below is a diagram describing the ECLSS cycle:

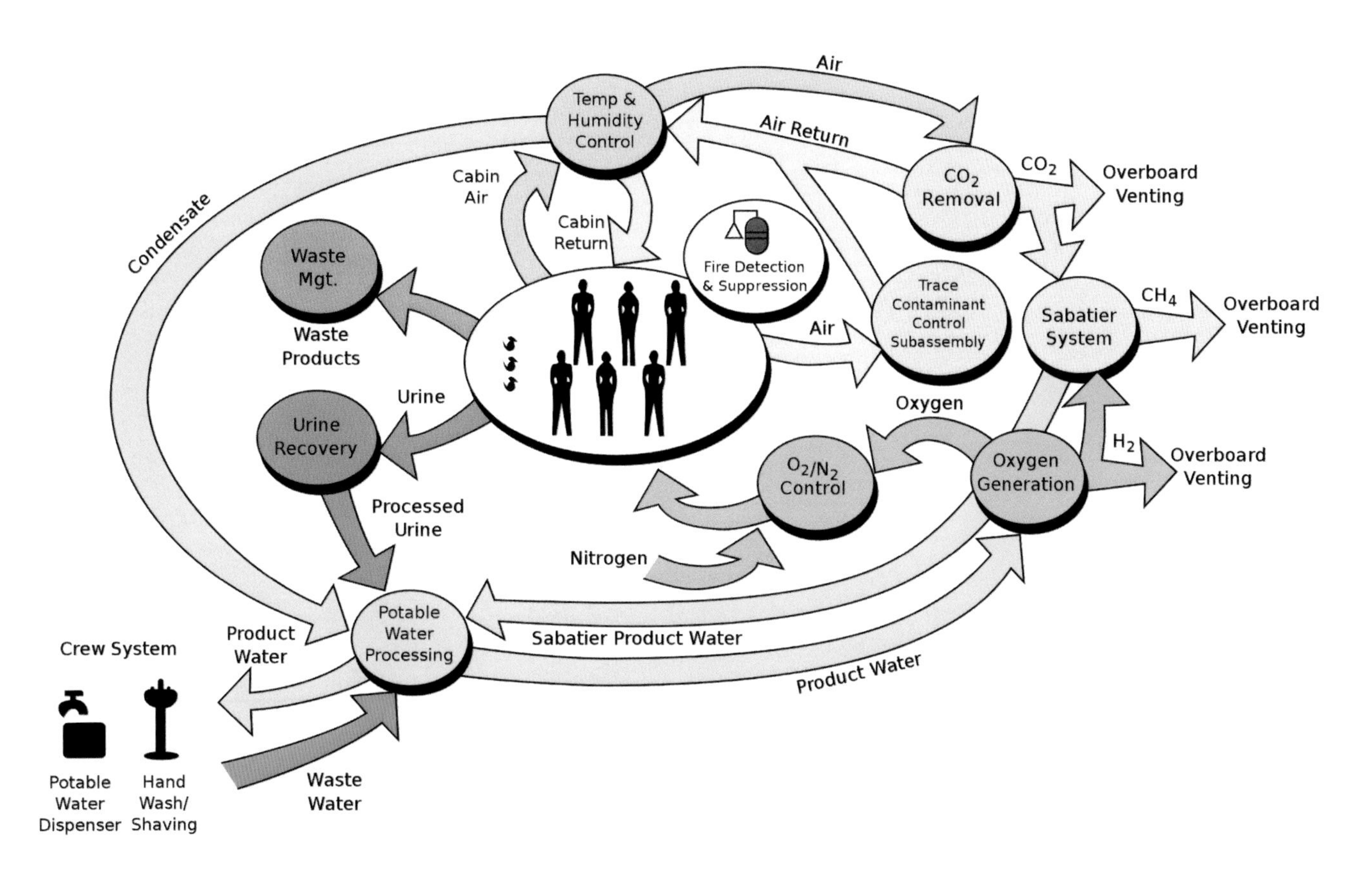

ECLSS system diagram. Source: NASA

The ECLSS is a good starting point on our journey to the next step for long term sustainability off-world. Some of the components, such as the famous urine to water distillation system[62], and the Sabatier Reaction system[63] for converting carbon dioxide + hydrogen into water for drinking and methane for fuel, could one day be used on Earth as well.

As NASA and its contractors Paragon Space Development Corporation[64] and Collins Aerospace[65] continue to perfect a nearly full cycle air and liquid processing system, this could be the foundation of larger space stations which could host many more plants, for both oxygen and for farming.

There must be a way to augment the mechanical or chemical life support systems with a natural life support system. The first reason is simplicity. Some plants have excellent filtration systems. The second reason is backups and redundancy. We need more redundancy in a life support system in case one or more systems fail. That includes designing small gardens that can filter the air inside every module of the space station! Finally, you need to humanize the off-world experience, and that includes having plants nearby to remind you of home. We are all Children of Earth, and we have a synergy with Earth's ecosystem that has kept humans alive since the beginning. Sorry if that sounds a bit woo-woo, but nature has the best life support system we know of.

Farms in Space

Several prototype space greenhouse systems are being tested on Earth. For example, the Prototype Lunar Greenhouse (LGH) at the University of Arizona's Controlled Environment Agriculture Center is designed as a Bioregenerative Life Support System (BLSS) via an innovative hydroponic plant growth chamber. Centered on using plants to sustain a continuous vegetarian diet for astronauts, the BLSS employs plants and crop production to grow food, provide air revitalization, manage water recycling, and manage waste recycling for the crew.[66]

---

62. "Pre-Treatment Solution for Water Recovery," *NASA Technology Transfer Program*, accessed October 25, 2022, https://technology.nasa.gov/patent/MSC-TOPS-68.
63. Christian Junaedi et al., "Compact and Lightweight Sabatier Reactor for Carbon Dioxide Reduction" (*41st International Conference on Environmental Systems*, Portland, OR, 2011), https://ntrs.nasa.gov/citations/20120016419.
64. "Environmental Control & Life Support Systems (ECLSS)," Paragon Space Development Corporation, accessed October 26, 2022, https://www.paragonsdc.com/what-we-do/life-support/environmental-control-life-support-systems-eclss/.
65. "Crewed Missions," *Collins Aerospace*, 2022, http://www.collinsaerospace.com/what-we-do/industries/space/crewed-missions.
66. David Story, "UA-CEAC Prototype Lunar Greenhouse," *University of Arizona, College of Agriculture and Life Sciences, Controlled Environment Agriculture Center*, 2010, https://www.ag.arizona.edu/lunargreenhouse/.

Concept: "Prototype Lunar Greenhouse (LGH)," Source: University of Arizona, Controlled Environment Agriculture Center, College of Agriculture and Life Sciences, 2010.

A homestead in space would be designed more like a greenhouse. Greenhouses on Earth keep the heat in, so that plants keep warm in the evenings and in the winter. Because the real estate inside the space station is at a premium, why not incorporate the living areas in the middle of the greenhouse? From a psychological standpoint, being able to look outside your window to see wide swaths of greenery is a pleasant experience. From a design and safety standpoint, having living quarters in the core of the station offers better protection from radiation (the plants and "water wall" cover the outer shell of the station) and the occasional micrometeoroid (the outer shell of a future space station would have multiple layers of metal "Whipple Shielding" + Kevlar fabric to reduce impacts).

Plant growing has been going through a lot of intense research with NASA. Onboard the ISS, the "Veggie" program has shown lots of success.[67] They have been very successful with growing fresh lettuce onboard the ISS. Hydroponic systems like Veggie seem to be very effective, as it requires the roots be attached to a water nutrient source. LED lights are used for sunlight simulation. After science tests are completed on the plants, some of them are consumed for the occasional meal. As the ISS is a research facility, there is no room to start a large garden or farm to support the entire crew. Designing a large space farm will become critical to survival as you develop a space homestead.

Let's talk about how you water those plants. Where do you store the water? Can water have secondary functions while it is waiting to be used to hydrate plants and humans?

---

67. Anna Heiney, "Growing Plants in Space," Text, *NASA, Kennedy Space Center*, July 12, 2021, http://www.nasa.gov/content/growing-plants-in-space.

Water Walls: Shielding and Sustenance

You really do not want to be outside the safety of the Earth's Magnetic field for too long, as the direct, continuous blast of solar and cosmic radiation can be fatal over time. As mentioned earlier in this chapter, it is equivalent to giving your body an X-Ray at the doctor's office nearly EVERY day. So what can you as a designer do about it? One answer, developed by engineers and space architects at NASA Ames Research Center, is the "Water Wall" concept:[68]

A 20 cm thick wall of water, stored in polyethylene bags, can absorb most of the radiation, due to the high count of hydrogen atoms in the water (H2O) and the polyethylene ((C2H4)n). Astronauts need a lot of water for drinking, washing, cleaning, food prep, and for watering plants. Large farms require lots of water in order to grow food. Hundreds of Polyethylene bags are needed to line the outer walls of a space station. These water bags can help block much of the radiation. In their design concept, algae in some of the bags can help process the waste water.

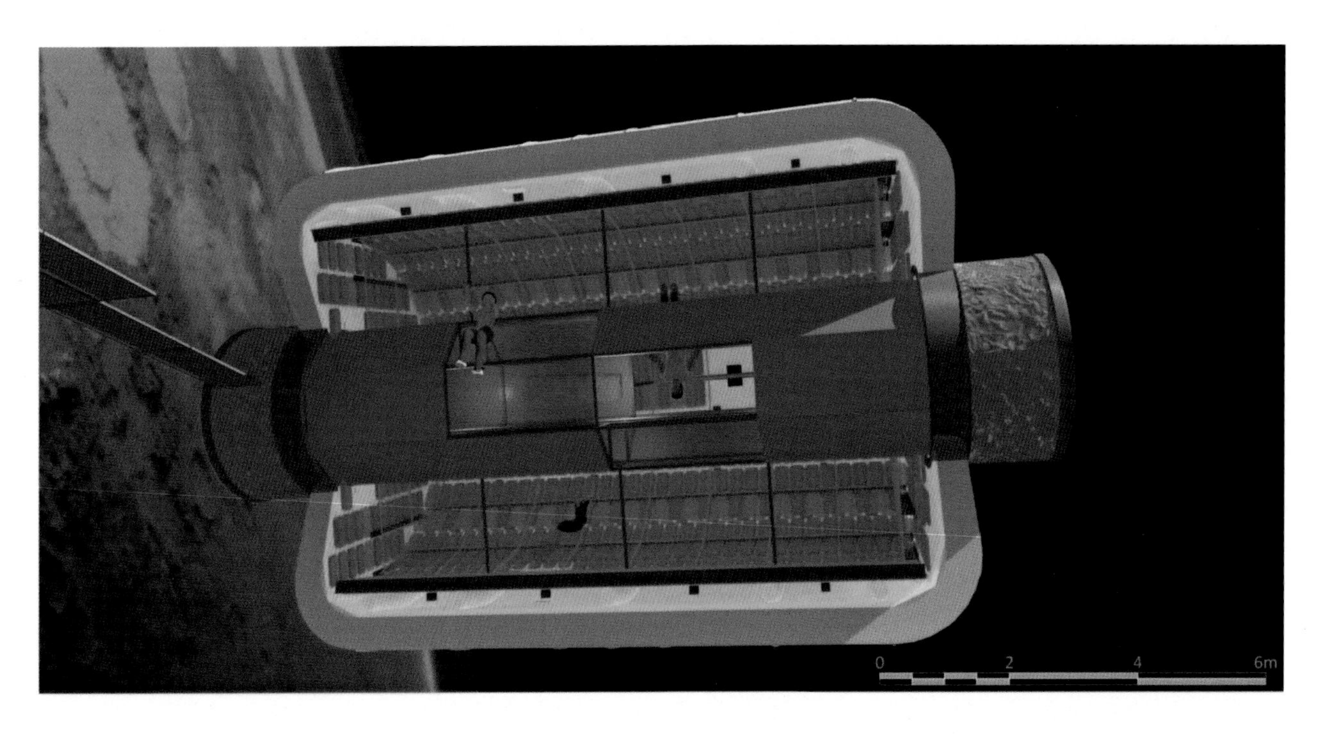

Cross-section of a Water Walls system installed in an inflatable structure. Individual bags are shown and pockets attached to an internal frame that is supported by a central truss. Source: Franois Levy for Dr. Marc Cohen, NASA Ames Research Center.

We have to look at what exists now for keeping astronauts alive, and then develop better, more reliable life support systems. While we are at it, let's design it to be simple, reliable, and sustainable.

Now that you understand the basics of survival in space, how do you make our new space station more like a home?

---

68. Michael T. Flynn et al., "Water Walls Architecture: Massively Redundant And Highly Reliable Life Support For Long Duration Exploration Missions," November 12, 2018, https://ntrs.nasa.gov/citations/20190001191.

The Case for Gravity

When developing this book, I assumed we are designing next-generation space stations based upon the traditional model: a Skylab or ISS-style space station without gravity, and that we have to find ways to adapt to the effects of microgravity on our bodies and our ways of living. That is still the case.

There is however, another option: designing a space station that creates artificial gravity. Sometimes called variable gravity space stations, these are the classic ring or cylinder designs that you've seen in science fiction movies. The idea goes all the way back to the space rocket visionary Konstantin Tsiolkovsky, who not only developed the rocket equation, also discussed the need for a rotating habitat to counteract the loss of muscle tone. Space pioneer Werner Von Braun proposed a ring station which, along with his rocket designs, inspired a generation of Americans in the 1950's.[69] You've probably seen the famous double-ringed space station in the science fiction movie "2001: A Space Odyssey." In the early 1970's, in the afterglow of NASA's successful Apollo Lunar Missions, many groups proposed building large orbital cities. Dr. Gerard K. O'Neill was famous for his "The High Frontier" book which proposed a series of giant spinning cylinders, which could hold thousands of people.[70] The NASA Stanford Torus Study proposed a massive ring space station with a supporting infrastructure of a lunar mining base and solar power satellites.[71]

All of these designs refer to a massive structure: something up to 200m (or more!) in diameter that can spin and create some semblance of an Earth normal environment. The massive size is required to reduce the dizzying effects of different gravity levels from your feet to your head. The ISS is modest in comparison: it is only 108 meters (357 feet) in length.

The challenge, back then, and to this day, is the cost of research, development, and actual construction. How do you build such a massive thing? One issue, of course, was the expense of sending anything like this into orbit. At the cost of $10,000 per pound of payload to orbit, the US Space Shuttle could not fly continuously enough to send all the components of an artificial gravity space station to orbit. With payload prices dropping, thanks to the advent of SpaceX's Falcon 9, Falcon Heavy, and the upcoming Starship, such large scale construction is now entering the realm of possibility.

The Competition

Three companies have announced they are planning to fly artificial gravity space stations: ABOVE: Space Development Corporation, GRAVITICS, and VAST Space.

---

69. Michael Neufeld, "Mars Project: Wernher von Braun as a Science-Fiction Writer," *National Air & Space Museum, Smithsonian Institution* (blog), January 22, 2021, https://airandspace.si.edu/stories/editorial/mars-project-wernher-von-braun-science-fiction-writer.

70. Gerard K. O'Neill, *The High Frontier: Human Colonies in Space*, 3rd Edition (Burlington, Ontario, Canada: Apogee Books, 2000), https://www.cgpublishing.com/Books/Highfrontier.html.

71. Richard D. Johnson and Charles Holbrow, "Space Settlements: A Design Study" (NASA Ames Research Center, January 1, 1977), https://ntrs.nasa.gov/citations/19770014162.

ABOVE's space station design consists of modules connected to a circular truss system. These modules are similar to NASA's TransHab[72] or Bigelow Aerospace's inflatable modules[73] that could spin fast enough to generate from 0 to 0.55 Gravity. Between each module will be an airlock to dock to various spacecraft, including some of Sierra Nevada Corporation's "Dream Chaser" spacecraft as escape pods.[74]

GRAVITICS's StarMax is big (400 m^3 cubic meters of usable volume). It has strong Micro-Meteoroid and Orbital Debris (MMOD) protection with its multi-layer walls.[75] StarMax also has body-mounted solar panels, and lots of windows.[76] It is not clear from currently available public documentation what rotational speeds it will achieve.[77]

Meanwhile, VAST Space's Haven-1 module reminds me so much of a small recreational vehicle, and that is a good thing.[78] It is a single, standalone space station, designed to work as an extension of the Dragon 2 space capsule and fit harmoniously with the rest of the SpaceX launch system (the team is made up of former SpaceX people). It will provide microgravity and Lunar artificial gravity environments. The company also has plans for a much larger 100-meter "spinning stick station," but the details are not publicly available.[79]

Benefits of Gravity

After reading this chapter, the benefits of gravity on a space station should be pretty obvious. It will eliminate several of the dangers that I talked about at the beginning of this chapter. Let's look at that chart again.

---

72. Horacio delaFuente et al., "TransHab: NASA's Large-Scale Inflatable Spacecraft" (2000 AIAA Space Inflatables Forum; Structures, Structural Dynamics, and Materials Conference, Atlanta, GA, 2000), 9, https://ntrs.nasa.gov/citations/20100042636.
73. Jeff Foust, "Bigelow Aerospace Transfers BEAM Space Station Module to NASA," SpaceNews, January 21, 2022, https://spacenews.com/bigelow-aerospace-transfers-beam-space-station-module-to-nasa/.
74. "ABOVE: Space Development Corporation," ABOVE: Space Development Corporation, accessed October 29, 2023, https://abovespace.com/.
75. Eric Christiansen, Dana Lear, and Jim Hyde, "Micro-Meteoroid and Orbital Debris (MMOD) Protection Overview" (NASA Johnson Space Center, October 17, 2018), https://ntrs.nasa.gov/citations/20190001193.
76. "Gravitics—Starmax," Gravitics Inc., accessed January 10, 2024, https://www.gravitics.com/starmax.
77. Aria Alamalhodaei, "Gravitics Raises $20M to Make the Essential Units for Living and Working in Space," *TechCrunch* (blog), November 17, 2022, https://techcrunch.com/2022/11/17/gravitics-space-stations/.
78. Kristy Hutchings, "NASA Partners with Long Beach's Vast Space on Artificial Gravity Station Development," *Press Telegram* (blog), June 16, 2023, https://www.presstelegram.com/2023/06/16/nasa-partners-with-long-beachs-vast-space-on-artificial-gravity-station-development/.
79. "Roadmap — VAST," Vast Space LLC, accessed January 10, 2024, https://www.vastspace.com/roadmap.

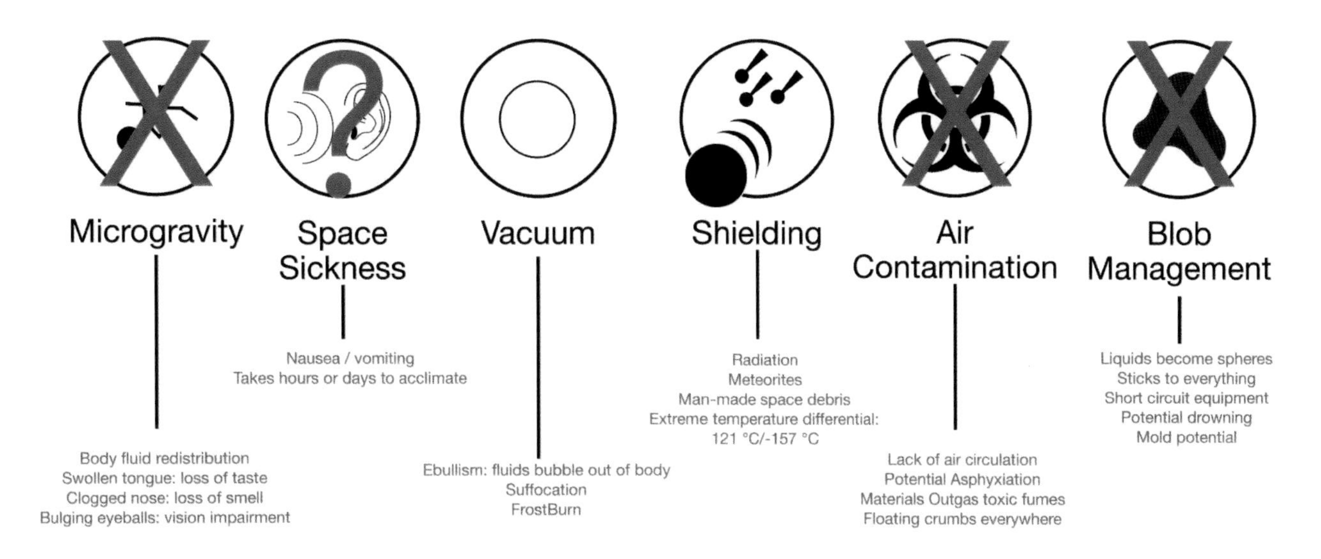

With an artificial gravity space station, many dangers to humans are reduced or even eliminated. Images by the author.

You would think this would be a no brainer. If a lot of the dangers from living in microgravity can be reduced or eliminated, then why not focus on artificial gravity space stations? The answer is... complicated.

Artificial Gravity: a Crash Course

Squeezing the essence of artificial gravity into a design book is nearly impossible to do without dropping lots of physics and calculus formulas onto your head. So instead I created visual images to help you. The goal here is to simply introduce you to the challenges of the concept. Thanks to the work of Dr. Theodore (Ted) Hall from the University of Michigan, whom I would consider the "Guru of Artificial Gravity," I present to you an abridged introduction to the topic.[80] Don't worry, I have links to many, many research papers and reports in the Bibliography for those of you who love to eat physics formulas for lunch.

Imagine a space station shaped like a donut, and with the spoke and hub elements of a wagon wheel. The whole station spins on its axis at about one revolution per second. The center hub spins slower than the outer ring of the station so that makes a logical point for a docking bay for spacecraft. Also at the hub is microgravity, so this is the best place for you to begin the transition from no gravity to 1G.

80. Theodore W. Hall, "Artificial Gravity," text (Theodore W. Hall, January 3, 2024), https://www.artificial-gravity.com/.

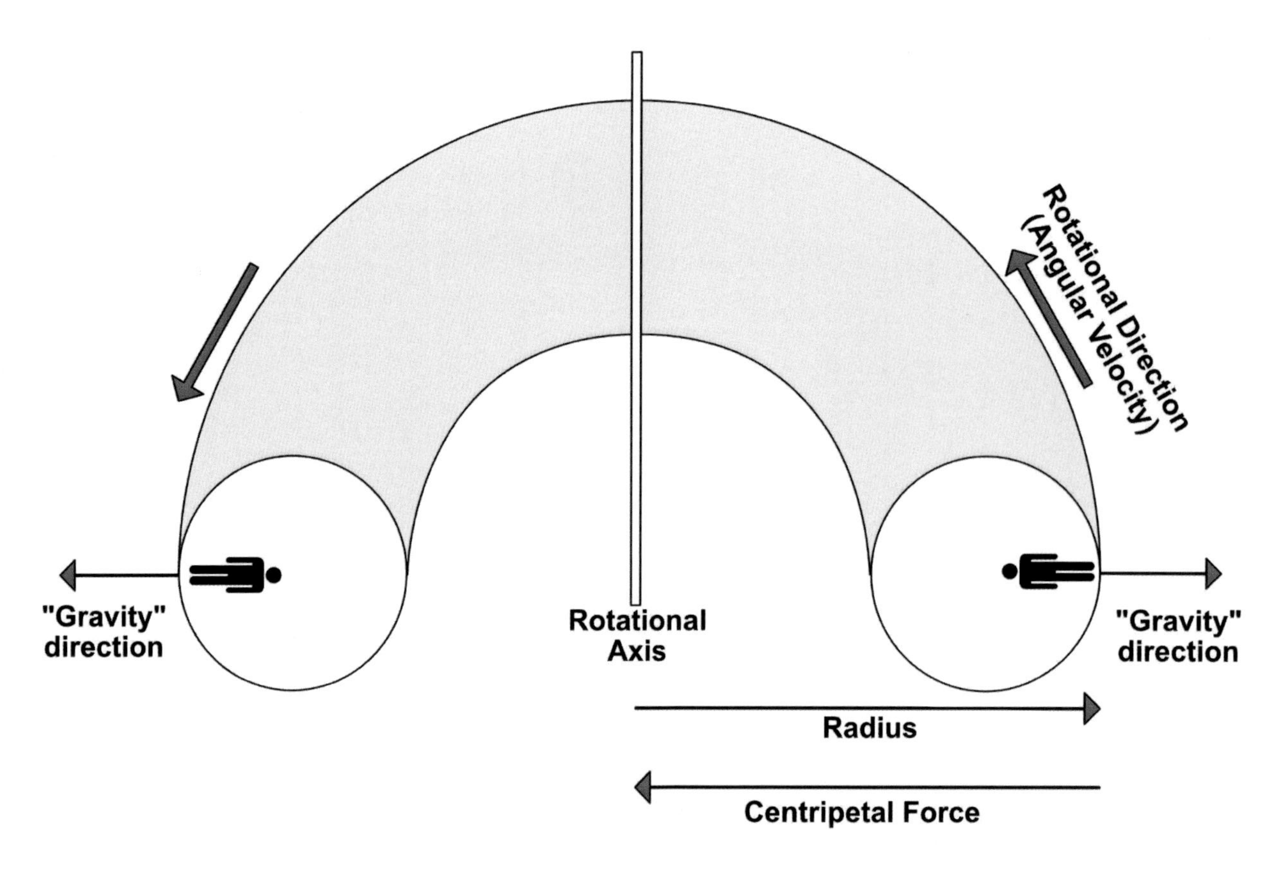

Cutaway view of ring space station concept. When the ring spins, it creates artificial gravity on the outer edge of the wheel. Image by the author.

So here's the hard part: even with the smallest wheel, you do get some semblance of gravity, where the centripetal force of the outer wheel rim creates the illusion of gravitational attraction for your feet. There is a catch: there is a thing called the gravity gradient. The closer you are to the center of the wheel, the less gravity there is. For example, that means if you are standing up with your feet on the wheel rim, the 1G you would feel on your feet gradually changes to half G in your midsection, to Zero G in your head. See the following illustration to get an idea of what I am talking about. The early tests in centrifuges by NASA and the Russian space agency in the 1970's proved how quickly people get sick in this situation.[81] [82] So to keep from getting sick, you need to increase the size of the wheel. The station wheel size needs to increase in order to decrease the gravity gradient between your head and toes.

81. J. A. Green and J. L. Peacock, "Effects of Simulated Artificial Gravity on Human Performance" (NASA Langley Research Center, November 1, 1972), https://ntrs.nasa.gov/citations/19730003384.

82. *Can The Human Body Handle Rotating Artificial Gravity?* (Scott Manley YouTube Channel, 2021), https://www.youtube.com/watch?v=nxeMoaxUpWk.

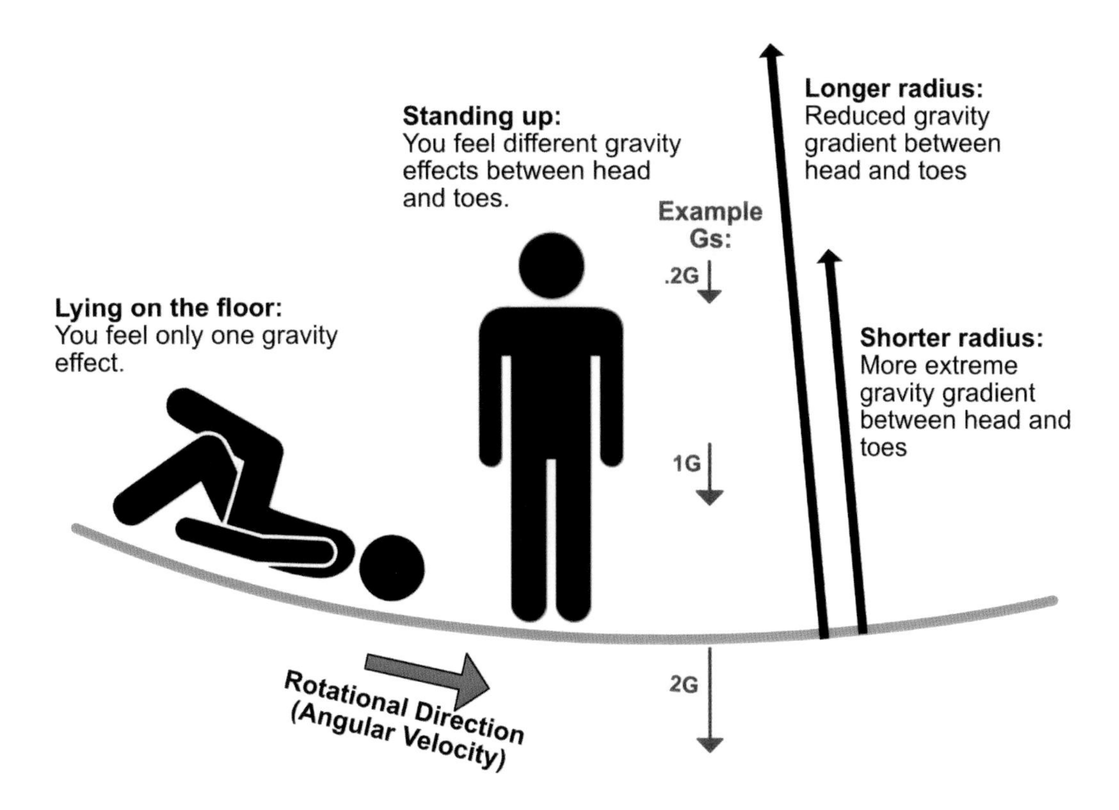

Gravity gradient visualization. Image by the author.

But wait, there's more! Coriolis forces come into play when you do things like turn your head in an orientation away from the spin of the station. Compare this to a roller coaster ride. Just as you're going down from the top of a hill and you turn your head to the right or left, there's a very good chance you're going to get sick. For military pilot training, there is this chair that the trainer sticks you in. When the chair spins, you have to place your head in different orientations. The faster you spin, the more likely you're going to get sick. Some people are more adaptable than others to this situation.[83]

Researcher Dr. Theodore Hall made a composite of 5 research studies that summarized the results of experiments with human subjects in centrifuges and rotating rooms on Earth.[84] Based on their calculations, humans can tolerate a rotating habitat going 1-2 revolutions per minute (RPM) if the rotational radius of the space station was 100 to 1000 meters, or 328 feet to 3280 feet, which equals 1 to 10 football fields long! That's just the radius length, which means it's only half the size of the entire station. By comparison, the ISS is just over one football field in length: 108 meters (357 feet) from end-to-end. Note that these are rough estimates based on earthbound studies. Until someone actually builds one of these giant spinning wheels in space, we will never know how accurate this data is.

---

83. "Barany Chair—Aerospace Medicine—AMST," AUSTRIA METALL SYSTEMTECHNIK (AMST) GmbH, accessed January 21, 2024, https://www.amst.co.at/aerospace-medicine/training-simulation-products/barany-chair/.

84. Theodore Hall, "Artificial Gravity Visualization, Empathy, and Design," in *Space 2006*, AIAA SPACE Forum, AIAA-2006-7321 (San Jose, CA: American Institute of Aeronautics and Astronautics, 2006), 22, https://doi.org/10.2514/6.2006-7321.

In general, the larger the space station, the less dizzying effects of Coriolis forces, and less of the gravity gradient will be upon your body.[85] Coriolis forces also affect how an object is dropped, depending upon the direction you throw it. This is a complicated discussion which you can pursue more in Dr. Hall's reports on his website (www.artificial-gravity.com).

Concept: Sleeping Ring

Since the benefits of an artificial gravity space station have yet to be proven out in space, there needs to be a way to start small. (Some experts yelled at me saying that "you cannot start small," but until some benefactor comes out of the blue with a billion dollars to build a massive O'Neill colony from scratch, we gotta try something). One concept that many of the researchers have not discussed is making what I call a "sleeping ring." What if you designed a bedroom in a spinning ring? If you can achieve something resembling 1G (or any percentage of 1G), for a person while their body is aligned flat on the floor while sleeping, this allows your body to recover from the effects of microgravity for part of the day. My idea is basically creating a hybrid space station: where the main work and play spaces are microgravity, but when you go to sleep you have a sleeping ring that counteracts some of the effects of microgravity.

Fortunately, someone else at NASA already proposed such a sleep ring.

There was a NASA proposal in 2011 for a transplanetary spacecraft for long term living, based on current technology. It used a lengthy acronym to fit the name: Nautilus-X (Non-Atmospheric Universal Transport Intended for Lengthy United States Exploration).[86] The concept has inflatable habitat elements including a ring structure to generate artificial gravity. The spacecraft's centrifuge is equipped with an external dynamic ring, with a flywheel to manage rotation. Basically, they are proposing to design and build the first practical interplanetary spacecraft! Although this concept never left the PowerPoint phase, it was thoroughly researched and the authors even proposed building a demonstration ring onboard the ISS.[87]

---

85. Theodore W. Hall, "Artificial Gravity in Theory and Practice" (International Conference on Environmental Systems (ICES), Vienna, Austria: 46th International Conference on Environmental Systems, 2016), 20, http://hdl.handle.net/2346/67587.

86. Edward M. Henderson and Mark L. Holderman, "Technology Applications That Support Space Exploration" (47th AIAA/ASME/SAE/ASEE Joint Propulsion Conference, San Diego, CA: AIAA, 2011), https://ntrs.nasa.gov/citations/20110013138.

87. Mark L. Holderman, "Nautilus-X: Multi-Mission Space Exploration Vehicle" (NASA Technology Applications Assessment Team, FISO Telecon presentation, January 26, 2011), https://web.archive.org/web/20110304044259/http://spirit.as.utexas.edu/~fiso/telecon/Holderman-Henderson_1-26-11/Holderman_1-26-11.ppt.

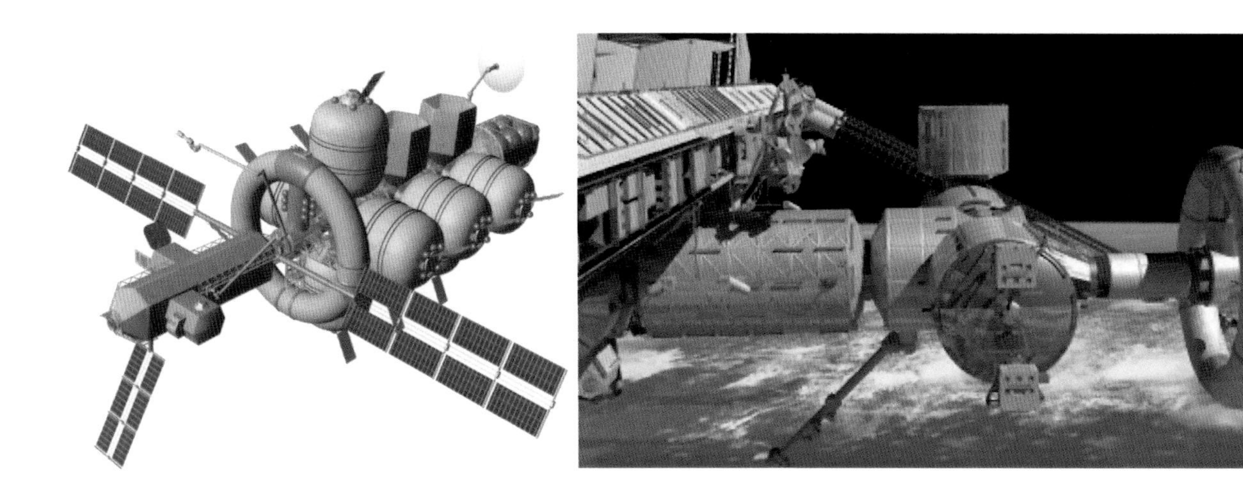

Left: NASA's Nautilus-X long duration space journey vehicle concept.
Right: Proposed ISS Centrifuge demo. Source: NASA.

Now that you've been exposed to the many design challenges of space travel, let's dive deeper into specific topics. First up: how are you gonna eat?

Chapter 2:

Food and Drink

Meat Sacks

When you go traveling, you typically buy or bring along snacks, food and drinks to keep you energized. Usually that means partaking at the airport or hotel restaurant, or the snackbar. What if you went to a place where not only the taste of the "usual" food changes, but it is also no longer sufficient to keep you alive?

In the Pixar animated movie "WALL-E," the human inhabitants of the starship AXIOM appear obese and barely able to walk.[88] While it is true they are living a decadent life, where everything is catered to them by robots, there is much more to their obesity. Their bodies have less bone mass. They are slowly turning into blobby meat sacks.[89]

This is what happens to astronauts in space. As mentioned above, you begin to lose calcium at a dramatic rate in a lower gravity environment. On average, astronauts lose about 1% of their bone mass for every month spent in space, which could make long term living in microgravity, or a trip to Mars, very challenging.[90]

One reason for this is adaptation. For thousands of years, humans and other animals developed strong bones and muscles to resist gravity: so that you can stand, walk, run or jump. People adapted to high altitudes, cold climates, and other extremes. What happens, though, when gravity disappears? No one really knew until the age of space travel began. For the first years of NASA's Mercury, Gemini, and Apollo programs, the missions were relatively short. From one day to maybe a week, astronauts were exposed to microgravity. NASA's doctors and scientists could observe subtle changes to the human body, but there was not enough time to determine long term effects.

---

88. "WALL-E," Pixar Animation Studios, accessed November 22, 2022, https://www.pixar.com/feature-films/walle.
89. *Fitless Humans (WALL·E)* (Pixar Animation Studio, 2013), https://www.youtube.com/watch?v=s-kdRdzxdZQ.
90. Canadian Space Agency, "What Happens to Bones in Space?," Canadian Space Agency, August 18, 2006, https://www.asc-csa.gc.ca/eng/astronauts/space-medicine/bones.asp.

All that changed with the age of the space stations. With Skylab, Salyut, MIR, and the ISS, the astronauts and cosmonauts spent many weeks or months in Earth's orbit. It was then that the debilitating effects of bone and muscle loss became apparent. As the saying goes, use it or lose it. Exercise routines became an important part of a space traveler's daily routine. It helped reduce the bone and muscle loss, but it was not enough.

Star Trek Replicator

In the classic TV show Star Trek, Captain Kirk goes to the food replicator, he orders his food, and magically it appears on a plate in front of him. How does that happen? If you want to build a starship with all the magical bells and whistles from a science-fiction show, then you better start working on it right now. Case in point: the food replicator.[91] If you were to imagine generating almost any type of food from scratch, how would you do it? There's been some amazing research on synthetic meats and other prepared foods appearing on the market today. So the idea of creating fake steaks is actually becoming real.

Lab Grown Foods

The precursors of replicator-made foods are being developed now. There are currently dozens of companies developing processes to grow chicken, fish and cow meat in the lab. Companies like Aleph Farms, Eat Just, Meatable, and UPSIDE Foods, for example, are pushing into new territories in cultivating animal cells and growing them in vats.[92] Before you scream "EWWWWW," think about it. The human population on Earth reached 8 Billion people in 2022.[93] We have been consuming the planet's resources almost to the limit, and generating tons of carbon and methane waste. By engineering meat in a lab, using real animal cells, we could reduce water consumption and carbon emissions dramatically. Along the way, we could prevent killing millions of animals per year...if someone can make the perfect hamburger!

---

91. "Replicator," Memory Alpha, November 27, 2023, https://memory-alpha.fandom.com/wiki/Replicator.
92. Polaris Market Research, "Global Cultured Meat Market Size Estimated to Reach USD 499.9 Million By 2030, With 16.2% CAGR: Statistics Report by Polaris Market Research," PR Newswire, accessed December 2, 2022, https://www.prnewswire.com/news-releases/global-cultured-meat-market-size-estimated-to-reach-usd-499-9-million-by-2030--with-16-2-cagr-statistics-report-by-polaris-market-research-301523708.html.
93. "World Population to Reach 8 Billion on 15 November 2022," United Nations, Department of Economic and Social Affairs (United Nations, November 15, 2022), https://www.un.org/en/desa/world-population-reach-8-billion-15-november-2022.

Meanwhile, vegetarians have been enjoying a plant-based meat renaissance with the popularity of many brands including The Better Meat Company, Impossible Foods and Beyond Meat.[94] On top of that, researchers are working on methods to grow fruits and fruiting vegetables separately from the supporting plant by developing the fruiting body in a test tube on a medium of sugars and nutrients.[95]

If you can grow food in these settings, how long will it take to build a machine that could grow and prepare you meals on Earth or in space?

A research team at University of California, Davis, have genetically modified a type of green lettuce to provide a human parathyroid hormone (PTH) to stimulate bone formation and could help restore bone mass in microgravity. It just means eating a very large salad every day: about 380 grams, or about 8 cups, of lettuce daily are needed to get a sufficient dose of the hormone. The research is in the early stages, and the lettuce still needs to be tested before it is safe to eat.[96]

Now hold on, you are probably saying, aren't GMO foods bad? Not necessarily. Plants and animals have been raised for centuries to feature one or another traits. A strawberry, for example, has many varieties, and the sweet ones are favored by farmers and consumers. Standard farming techniques can filter the sweet from the not so sweet ones. A truly Genetically Modified plant is chemically altered in a lab, usually to resist disease, and then grown to replace older crops that have been devastated by disease.[97]

Why do foods have to be genetically modified for use in space? For the most part, no, they do not need to be. Genetic modification is another tool to deal with the effects of microgravity and radiation. You could simply require the space traveler to take lots of pills and shots every day to help augment the required exercises, but that assumes you can guarantee regular, consistent resupply ships coming to the space station. An alternative is to grow your own medicine in the form of food plants.

With that little teaser, let's talk a little history and the current state of food in space.

---

94. Katie Bandurski, "19 Plant-Based Meat Brands Every Vegetarian Needs to Know," *Taste of Home* (blog), November 15, 2022, https://www.tasteofhome.com/collection/vegetarian-brand-names/.

95. Ramin Ebrahimnejad, "Lab-Grown Fruits," *Association for Vertical Farming* (blog), January 15, 2021, https://vertical-farming.net/blog/2021/01/15/fruit-of-knowledge/.

96. "Space-Grown Lettuce Could Help Astronauts Avoid Bone Loss," American Chemical Society, March 22, 2022, https://www.acs.org/content/acs/en/pressroom/newsreleases/2022/march/space-grown-lettuce-could-help-astronauts-avoid-bone-loss.html.

97. "Food, Genetically Modified," World Health Organization, accessed November 23, 2022, https://www.who.int/health-topics/food-genetically-modified.

From Cubed Food to Cuisine

Bags of Space Station food and utensils on a tray. Note the amount of single-use plastic packaging. Source: NASA.[98]

> *"Early US space food was highly engineered to minimize mass and volume and to prevent any possibility of food scraps contaminating the small cabins of the early NASA spacecraft. Space menu items consisted primarily of puréed foods in squeeze tubes, small cubed food items coated with an edible film to prevent crumbs, and freeze dried powdered food items. It was agreed by most that early space food was, to say the least, unappetizing."*
> *–Charles T. Bourland*

This quote comes from Charles T. Bourland and Gregory L. Vogt's enlightening book, "The Astronaut's Cookbook." Mr. Bourland spent 30 years at NASA Johnson Space Center's Space Food Laboratory developing food and food packages for spaceflight. His book covers the nutritional, technical,

98. Amiko Nevills, "NASA—Space Food Laboratory Gallery," NASA Johnson Space Center, November 25, 2007, https://www.nasa.gov/audience/formedia/presskits/spacefood/gallery_jsc2003e63872.html.

psychological, and even political challenges of food and drink for NASA astronauts.[99] I tried to summarize key details from the book below to help you understand our current space food system so that we can figure out a better way of sustaining humans off-world.

Size and Weight Limitations

As NASA matured its space program, so did the food systems. Due to restrictions in size and weight, concentrated foods were emphasized. The real challenge for space food became apparent when astronauts were returning from early space flights with decreased body weight, with food uneaten. How do you increase the flavor, variety, and quality of food so that the astronauts actually want to eat it?

The only space stations to have a dedicated refrigerator and freezer for people were the US Skylab and the Soviet/Russian MIR space station. The US Space Shuttle was too small and only did short trips (up to two weeks).

It was only in 2020 that a dedicated refrigerator/freezer for human food was installed onboard the International Space Station. The various other freezers were dedicated to preserving science experiments. Once again, the ISS is a research lab and human comforts are not a priority. That attitude is slowly changing, as NASA astronauts stay longer and longer on the space station. For example, ice cream is in high demand, but so far the astronauts have to settle for the freeze-dried kind.[100]

Now that you have a historical perspective, what kind of foods are safe for the space traveler?

Variety is the spice of life

I get bored easily. Sometimes I have the same type of food for a while and then just go crazy and just get something radically different. The same could be said with my interest in teas. On any given day, I'll switch from Vanilla Almond tea, to Cardamom tea, to Paris tea, to Earl Grey Tea (hey, I like tea ok?). Imagine having to live on an airplane for a year, and all you had to eat was the same meals for breakfast, lunch, and dinner? For an entire year? That would drive anyone crazy. This is another factor to consider when you are designing cuisine for space travel.

Space Food Types

Remember, the most critical factor for food in space travel is the weight of the cargo. Current space rockets have an extremely limited carrying capacity. That means each and every item is evaluated and ways are found to make it lighter. Future space rockets, such as the SpaceX Starship, will dramatically increase cargo capacity, but weight will still need to be kept under control.

---

99. C.T. Bourland and G.L. Vogt, *The Astronaut's Cookbook*, 1st ed. (New York: Springer Science + Business Media, LLC., 2010), https://link.springer.com/book/10.1007/978-1-4419-0624-3.

100. Shi En Kim, "The Quest to Build a Functional, Energy-Efficient Refrigerator That Works in Space," Smithsonian Magazine, July 27, 2021, https://www.smithsonianmag.com/innovation/quest-to-build-functional-energy-efficient-refrigerator-that-works-in-space-180978281/.

The second most important thing is shelf life. Food on the ISS needs to last for at least three months at a time, preferably up to three years! On the short term end, one great creature comfort astronauts get is fresh fruit and vegetables! The astronauts and cosmonauts onboard ISS are very excited whenever a resupply ship arrives. NASA usually adds a few surprise goodies for them to enjoy, such as fresh fruit. When they arrive via a cargo ship, they need to be eaten immediately, as they will not stay fresh for more than a day or two. Also, the stench of decaying food will fill the space station with a nasty odor and it is hard to filter out (where would it go?)

If you have done any kind of camping, hiking, or extended road trips of any kind, some items in the following list will be familiar to you. This is the official NASA list of food types from "The Astronaut's Cookbook," and they help meet the requirements for low travel weight and long lasting flavor.

- **Rehydratable Food:** A great way to extend shelf life and make food lighter is to remove the water from it. In earlier spacecraft, such as Apollo and the US Space Shuttle, water was generated onboard by fuel cells. Fuel cells generate electricity for the spacecraft by merging oxygen and hydrogen. Water is the byproduct of this reaction, so that means you do not have to carry it to space. You can use the newly-generated water for drinking and re-hydrating the food.
- **Thermostabilized Food:** This is your typical canned food such as beans, tomatoes, peas, or artichokes. These foods are heated to destroy dangerous microorganisms. Since cans take up too much weight and mass, they need to be repackaged for spaceflight in flexible plastic pouches. The infamous military MREs (Meals Ready to Eat) fit in this category.
- **Intermediate Moisture Foods:** These foods have restricted amounts of water in them to keep them soft and moist. Examples include dried fruit and beef.
- **Natural Form Foods:** Nuts, granola bars, candy, and cookies are simply packaged and ready to eat without any fancy preparation.
- **Irradiated Meat:** Zapped with ionizing radiation to ensure long shelf life, these meats include smoked turkey, fajitas, beefsteak, and breakfast sausage.
- **Condiments:** These are your standard condiment packets you would see in a take-away restaurant or supermarket. Some flavors include hot sauce, mayonnaise, mustard, and ketchup. Salt and pepper, if dispensed from a paper packet or shaker, would make a giant floating mess. Instead, astronauts use eye-droppers. The pepper is suspended in oil, and the salt is suspended in water. You would take the dropper and press onto the food and squeeze. Note: it is hard to evenly dispense salt and pepper this way and you may get oversaturated spots.

Personally, I have used soy sauce as an excellent "liquid salt" and it does a great job of spreading the flavor all over hard boiled eggs and other foods. I would assume you could put soy sauce in an eye dropper and get the same effect in a space meal.

Playing with your food

Creature Comfort #1: Fresh baked cookies

In 2019, the ISS astronauts baked cookies in space for the first time! They were baked in a prototype oven built by NanoRacks, a small space business coordinator for NASA, and Zero G Kitchen, which creates appliances for microgravity use in long-duration space flights. The cookie dough was donated by DoubleTree Hotel (famous for their yummy chocolate chip cookies they give to hotel guests).[101] Each of the five cookies were in a special sealed pouch, which were baked individually at different times at 150°C (300°F). The fourth and fifth cookies—one baked for 120 minutes and left to cool for 25 minutes, and the other baked for 130 minutes and left to cool for 10 minutes—were deemed to be the most successful. Sadly, none of the cookies were actually eaten, because the crumbs would fly everywhere! They were part of a science experiment and were sent back to Earth. More research needs to be done on different types of baking and cooking ovens that work in a microgravity environment. The potential is quite good here.

If you have very young kids, one of the things you constantly teach them is to not play with their food. When you are in space however, playing with food takes on a whole new dimension, literally. There are many many videos on the Internet showing the astronauts and cosmonauts sucking giant blobs of orange juice in mid-air, attacking floating M&M candies, or tossing tortillas across the room onboard the ISS. The food, just like the astronauts, are floating (technically falling and missing the ground).[102]

---

101. Jason Daley, "With a 'Zero G' Oven, Astronauts Can Have Their Cookies, but They Can't Eat Them Too," Smithsonian Magazine, November 5, 2019, https://www.smithsonianmag.com/smart-news/space-no-one-can-hear-you-nom-space-station-getting-cookie-baking-oven-1-180973455/.

102. *How to Prepare (Thanksgiving) Food in Space* (NASA Johnson Space Center, 2015), https://www.youtube.com/watch?v=60fxGvNLFtY.

As you can see below, lunch can become a very messy business!

Floating tomatoes and other food items. Flight Engineers Cosmonaut Sergei Treschev (left—reaches for hamburger which is floating) and Astronaut Peggy Whitson (right—holds packet) stand around table in the galley area in the Service Module (SM)/Zvezda. Photo taken during Expedition Five on the International Space Station (ISS). Source: NASA.[103]

103. Amiko Nevills, "NASA—Food in Space Gallery," NASA Johnson Space Center, November 25, 2007, https://www.nasa.gov/audience/formedia/presskits/spacefood/gallery_iss005e16310.html.

Some Assembly Required

Creature Comfort #2: Hot Meals/Fresh Food

Food will be the next revolution in space development. Currently on the International Space Station, astronauts use a glorified hot plate inside a suitcase to warm up food that is hermetically sealed in plastic bags. (The Space Oven mentioned above is a definite improvement). There is a food prep table, where the utensils and food components stick to the table with velcro or straps. Without restraints, items float away quite easily. As mentioned earlier, food crumbs just float right in front of your face while you are eating! If you are not careful, the crumbs have the potential to float into sensitive equipment, or even your nose! Regular bread is not sent up to the ISS due to the excessive debris that would fly around. That is why they used tortilla bread instead. Rice, especially sticky rice from Japan, is useful because it does not float away as easily as other types of carbohydrates. So as you can see, a number of constraints affect food preparation in space.

There is a great video where "Mythbusters" TV show stars Jamie Hyneman and Adam Savage interview astronaut Chris Hadfield about cooking and eating in space. This was part of Adam's TESTED Youtube channel.[104] You learn that astronauts may not have access to fresh fruit and vegetables for months at a time. Also on the show is their friend Chef Traci Des Jardins, who goes to the NASA Johnson Space Center Food Laboratory and taste tests the different foods that are rehydrated and processed for the astronauts. With the food limitations known, Chef Traci creates a steak and beans burrito for astronaut Hadfield. You get to see the challenge of just trying to assemble a simple meal. Every package, every food item, just wants to float or twirl away! The astronauts have a meal table that consists of velcro strips and elastic straps. Each food package and utensil has a piece of velcro on them to attach to the table, or they snug the item under a strap. Creating a meal with multiple components can be challenging. Just like in camping, food prep in space can be a complicated mess. That's why most campers simply like to be able to open a can of mini ravioli and eat the contents.

Eating Dinner

As mentioned earlier in this chapter, eating meals can be a multi-dimensional adventure! That is why NASA has almost all of its foods containerized in some type of plastic packaging. I know it seems kind

104. *Astronaut Chris Hadfield and Chef Traci Des Jardins Make a Space Burrito* (Adam Savage's Tested: YouTube, 2013), https://www.youtube.com/watch?v=f8-UKqGZ_hs.

of wasteful, but it's extremely convenient to do it that way. For future long-term living off-world, there may be ways to recycle some of the plastic, as in using them as containers for other foods. Many of the utensils we use today can still be used in space. Chopsticks can be very useful for grabbing stuff in mid air. The concept of a proper dining table may not work in a weightless environment. Everything needs to be strapped down before it floats away. So yes, lots of research still needs to be done to improve the art of eating without gravity.

Swollen Tongues

Astronauts love spicy foods, in part because the fluids flowing to their heads swell their tongues and reduce the taste bud senses. This swelling of the tongue basically means all foods taste different in space.

So now you have another challenge: how to adapt food flavors for space travelers? What tastes good on Earth may taste bland in microgravity. Heck, there is even research that on commercial aircraft people's taste buds change during high altitude flights.[105]

Story time: when I was living in Venice Beach, California, I had my first exposure to a Mexican fruit stand. This is a very common thing in California, where vendors sell all sorts of stuff from mobile carts wandering through the streets. I typically don't partake, but on one particular hot day, I thought, what the heck. So I asked a vendor for a bowl of fresh fruit. The gentleman started assembling the fruit and proceeded to put tons of salt and chili pepper onto it. The fruit flavor exploded in my mouth. It was amazing. I've never been a fan of spicy food culture, but I know a lot of people who really dig fiery hot sauces. Interestingly, this is a common solution to the bland food problem in space. (By the way, tortillas are commonly used instead of bread, for making sandwiches in space. The reason being is they don't flake away like regular bread does.)

Creature Comfort #3: Pan fried bacon

Bacon has several challenges. Even if you could cook it on a hot plate, it is extremely messy to clean up in microgravity. A ball of floating hot grease is not a fun thing to encounter. Another factor is the smell. Yes, the smell will be quite wondrous, but it may never go away. As mentioned in the first chapter this persistent smell is due to the lack of air circulation in microgravity. Maybe designers need to come up with a way to gather smells and send them out through the airlock?

105. Stacey Leasca, "5 Things to Know Before Ordering Food and Drinks on the Plane," Food & Wine, December 6, 2022, https://www.foodandwine.com/food-drink-airplane-ordering-tips-6836206.

Scoville Pepper Test

Someone needs to do a new type of flavor experiment in space. There's a thing called the Scoville scale.[106] It is used for measuring the flavor extremes of different types of chili peppers. Some people find this an enjoyable challenge. For offworld food prep, I suggest that someone needs to do the following experiment: have astronauts taste different levels of chili pepper seasoning (or the peppers themselves) on Earth, and then try them again onboard the ISS. We need to figure out what Scoville level they can taste and which ones they cannot due to the fluid swelling in their tongue and nasal cavity. The results of this test could help determine how much more spicy flavors are needed for foods to make them enjoyable for astronauts off-world.

So many other flavors exist besides peppers, and food scientists are working to create tests to detect levels of sensitivity to smoky, sweet, savory (which is usually considered the same as umami), and other flavors. As we move towards a space-bound future, food scientists have the opportunity to create objective methods for comparing earthbound versus off-world levels of flavor tolerance in human smell and taste.

Saturday Morning Science

NASA Astronaut Donald Pettit, during his classic "Saturday Morning Science" programs he ran while onboard the International Space Station during the early 2000's, demonstrated many unique features of liquids in a microgravity environment.[107] For example, he showed that you can drink tea using chopsticks![108] A liquid becomes a sphere when there is no gravity, because the surface tension is so strong that it wraps around itself! That surface tension, in turn, tends to be very sticky. Dr. Pettit simply placed the chopsticks on the outer edges of the liquid sphere, where it immediately latched onto the sticks!

Dr. Pettit conducted a whole series of experiments showing the unusual powers of surface tension and capillary action. These features are visible on Earth, but are usually overwhelmed by the force of gravity. For example a water droplet, what you see typically on a leaf or on a table, appears kind of flat or squished. When gravity is removed, that water droplet becomes more of a pure sphere. The cohesive or stickiness factor of the liquid becomes much more apparent.

The Art of Blob Management

It was Donald Pettit's work with liquids that blew my mind. His research culminated in an invention: the microgravity coffee cup. As mentioned before, astronauts drink liquids from squeeze bags with

106. "Chili Pepper Scoville Scale | Scovillescale.Org," *Scoville Scale* (blog), accessed November 27, 2023, https://scovillescale.org/chili-pepper-scoville-scale/.

107. Tony Phillips and Donald Pettit, "Saturday Morning Science | Science Mission Directorate," NASA Science, February 25, 2003, https://science.nasa.gov/science-news/science-at-nasa/2003/25feb_nosoap.

108. *Saturday Morning Science: Drinking Tea with Chopsticks in Microgravity Onboard the International Space Station.* (International Space Station, 2006), https://www.youtube.com/watch?v=7obLT4s2-HA.

tubes. It is technically a very safe and reliable way of dispensing drinks without making a mess in a weightless environment, but it takes away the enjoyment. As Dr. Pettit mentioned in one of his many interviews, the sense of smell is 40% to 60% of the enjoyment of a drink of coffee. So he decided to fix this problem.[109]

Clockwise: The first prototype of the microgravity coffee cup where it shows liquid flowing up on the narrowest point of a teardrop-shaped container.[110] The evolution of the microgravity coffee cup from prototype to 3-D print model to ceramic cup.[111] Astronaut Samantha Christoforetti drinking from the microgravity coffee cup while looking out the cupola window on the International Space Station.[112] Source: NASA.

---

109. Jeff and Dustin, "Fueled By Death Cast Ep. 106—DONALD PETTIT," Video, Death Wish Coffee Video Podcast: Fueled by Death Cast, accessed November 24, 2022, https://www.deathwishcoffee.com/pages/fbdc-ep-106-donald-pettit.

110. *Astronaut Demos Drinking Coffee in Space*, CollectSpace (International Space Station, 2008), https://www.youtube.com/watch?v=pk7LcugO3zg.

111. *Astronaut Donald Pettit on the Evolution of the Zero G Coffee Cup* (Death Wish Coffee Company: Fueled by Death Cast, 2018), https://www.youtube.com/watch?v=ugQIivUuuXk.

112. "Space Station Espresso Cups: Strong Coffee Yields Stronger Science—A Lab Aloft (International Space Station Research)," Blog, NASA Blog: A Lab Aloft (International Space Station Research), May 1, 2015, https://blogs.nasa.gov/ISS_Science_Blog/2015/05/01/space-station-espresso-cups-strong-coffee-yields-stronger-science/.

Dr. Pettit filed the first patent for a space-generated invention, which could not be created anywhere else except in space.[113] Dr. Pettit discovered it through play. By simply observing the interactions of liquids on different surfaces in weightlessness, it allowed him to develop a new thing. This creation was a huge leap in the realm of creature comforts.

I began thinking of creature comforts in space in terms of "blob management." How do you improve farming, showers, washing hands, heating and cooling, cleaning messes, and any other activity in space that deals with liquids in any form? The key was to understand how blobs work in microgravity. A "blob" is a generic term I made up to cover water, oils, alcohol, waste water, and other liquids that you may encounter through life off-world, how they act differently, and how to process them.

Creature Comfort #4: Eye Drops

They are useful, but as mentioned before with surface tension, if you are not careful with the application, the blob of liquid may coat your face.

Mad Science Projects

Back around 2005, I worked with the Space Tourism Society, a non-profit organization whose goal was to promote opportunities for more private space travelers to visit the ISS and eventually future space hotels (we call them Orbital Super Yachts and Orbital Cruise Ships).[114] I was fascinated with the design aspect of the Space Lifestyle. During that time I developed about a half a dozen different concepts, each one related to comfort, convenience and style. Below are some examples.

113. Mary Robinette Kowal, "The Need for Caffeine Was the Mother of Invention," *The New York Times*, November 2, 2020, sec. Science, https://www.nytimes.com/2020/11/02/science/space-station-coffee.html.

114. "Space Tourism Society | Welcome," Space Tourism Society, accessed December 18, 2022, https://spacetourismsociety.org/.

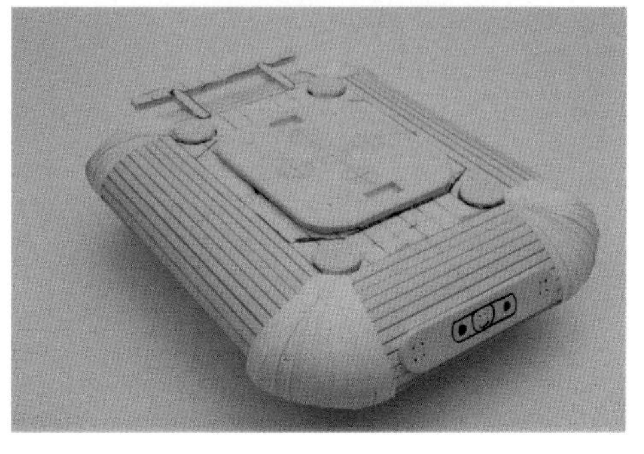

Clockwise, Left to Right: The Inflatable "G-Chair" for socializing (your body naturally goes into a fetal position in microgravity, so your knees hold you in place. The "Microgravity Cocktail Glass" (original prototype, before learning about surface tension and capillary action). The "Snuggle Tunnel" (for intimate activities in weightless spaces). The "Service Robot" (multipurpose, intelligent bot for cleaning, serving drinks, toxic gas/radiation detection, other functions). Source: the author.

Starting in 2009, for fun I worked with artist and engineer friends to build cocktail-making robots, also known as drinkbots. (This was back in the day when there was an underground art scene in San Francisco which overlapped the Burning Man scene which overlapped the Silicon Valley engineering/programming scene. It's a long story for another book). In the process of developing a drinkbot, I learned a lot about capillary action and fluid flow. I researched and tested gravity fed pumps, pressure fed pumps, and eventually settled upon the awesome peristaltic pumps. These pumps use different means of transferring liquids from point A to point B. My friends and I discovered that liquids are not very cooperative. They tend to go in their own direction. The challenge we had was getting precise amounts of liquids out of the bottle and into a mixer, then into the glass. Over the years, we developed slightly different processes with three different drinkbots, creating drinks in a variety of unusual ways.[115]

115. Samuel Coniglio, "DRINKBOTS," Obtainium Works, accessed November 24, 2022, https://www.obtainiumworks.net/drinkbots.

COSMOBOT:
Launching cosmic cocktails!

The Tea Engine:
Serving quality tea with class!

TIKITRON:
Make a sacrifice to the volcano deity, and receive a tasty tropical drink!

Left to right: "COSMOBOT" launches cosmic cocktails such as a Cosmopolitan. "Tea Engine" serves quality tea with class. The "TIKITRON" is a volcano deity that requires a sacrifice in order for you to receive a tasty tropical drink! Source: the author.

Each robot has a unique artistic theme: a Cosmic cocktail making rocketship,[116] a tea making robot, and an epic volcano deity which dispenses quality Tiki Cocktails when a person drops a wooden idol (a "sacrifice") into the top of a volcano![117] The theatrics and decorations hid the pumps, tubing, RFID tags, power supplies, and bottles from the audience. Behind the scenes, we used Arduino or Raspberry Pi microcontrollers to program and manage the pumps and dispensing of fluids. The pumps themselves evolved from pressure-fed to gravity-fed to peristaltic pumps. The advantage of peristaltic pumps is that they pull the liquids through the tubes no matter what the gravity situation is.[118]

---

116. *Robot Bartenders of BarBot 2013 Serve Up Drinks* (SOMArts Art Gallery: BarBot 2013, 2013), https://www.youtube.com/watch?v=zQaZ_-i0EJM.

117. *Cocktail Robotics Grand Challenge, 2016* (DNA Lounge, 2016), https://www.youtube.com/watch?v=Gcn8J7nHX3k.

118. "How Does A Peristaltic Pump Work?," TapFlow Pumps UK, accessed November 7, 2023, https://www.tapflopumps.co.uk/blog/peristaltic-pump-guide/.

Drinks in Space

Creature Comfort #5: Cold beer and cocktails

Absolutely, but within reason. Currently this is a taboo topic on the ISS, and NASA officially forbids it. Unofficially, several sources confirm that, yes indeed, drinking alcohol in space does happen. A handful of studies have shown that drinking small amounts of alcohol can reduce tension and improve relaxation[119] For social situations, drinking is acceptable, with a few caveats. Onboard the government-run International Space Station everyone onboard is trained to deal with crises that could happen at any moment, and there is legitimate concern that a drunk crew is a very bad idea. This is one reason why future space station designs need to be a lot more "idiot proof" for non-professional astronauts.

Then in 2014, after discovering astronaut Donald Pettit's work, I returned to one of my space tourism concepts, a cocktail glass. I worked with top notch designer friends Nick Donaldson and Brent Heyning to create a working 3D printed model of a microgravity Cocktail Glass. Inspired by astronaut Dr. Pettit's microgravity coffee cup, we dived deep into how surface tension and capillary action works and how liquids act differently in microgravity. The result was an amazing object: The Zero Gravity Cocktail Glass!

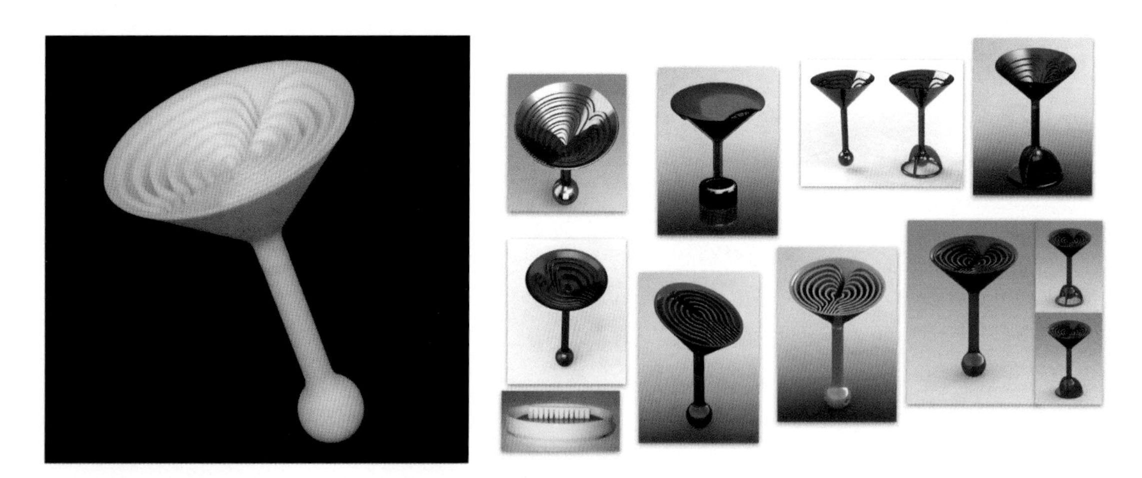

The Zero Gravity Cocktail Glass. Left to right: 3D printed glass by Nick Donaldson, Glass design evolution montage by the author (3D renders by Nick Donaldson). Source: the author.

119. Michael A. Sayette, "Does Drinking Reduce Stress?," *Alcohol Research & Health* 23, no. 4 (1999): 250–55, https://www.ncbi.nlm.nih.gov/pmc/articles/PMC6760384/.

For a brief time, we attempted to both patent and commercialize the glass by starting a company, with seed funding from celebrity bartender Russell Davis, and from my father-in-law. Even though we had excellent designers, we were woefully underfunded, and a bit naive of the business acumen needed to market this novelty object. Also we were a bit too obsessed with 3D printing, and learned the difficulty of manufacturing such complicated objects to a large scale. Finally, we were stumped by NASA's Public Affairs Office which officially forbids anything alcohol-related to fly into space. My friend Chris Carberry goes in-depth discussing this taboo in his book, *Alcohol in Space,* as others have encountered similar barriers when trying to send anything booze-related to space.[120]

Capillary action (whereby liquids naturally flow up narrow channels like straws), and surface tension (where a liquid's inherent "stickiness" would cause it to attach to things or itself) in space fascinates me. During the development of the Zero Gravity Cocktail Glass, we had the chance to interview Dr. Mark Weislogel, the scientist who does heavy-duty research on fluid dynamics, and helped refine the design of the zero gravity coffee cup with astronaut Donald Pettit. Understanding fluid flow in microgravity and designing plumbing that can manage it via capillary action has powerful implications for any system which manages fluids, such as kitchen sinks, toilets, farming, and life support.[121]

I consider these concepts critical to improving creature comforts in space. By understanding how blobs work, you can create devices to manage them better.

Then in 2015, an actual coffee maker was sent to the International Space Station!

The Art of Making Coffee

Creature Comfort #6: Coffee/Tea

Most people can agree that a fresh cup of coffee or tea can be quite relaxing mentally and physically. Many cultures worldwide place a special significance on making and sharing tea or coffee. However, on the International Space Station, all you get is a plastic bag, pre loaded with coffee or tea, and you use a device to inject hot water into it. You would simply drink the contents from a straw. You cannot smell or taste the flavor, because of the sealed nature of the bag, and of bodily fluids clogging your head.

120. Chris Carberry, *Alcohol in Space*, 1st ed. (McFarland & Company, 2019), https://mcfarlandbooks.com/product/alcohol-in-space/.

121. M. M. Weislogel et al., "How Advances in Low-g Plumbing Enable Space Exploration," *Npj Microgravity* 8, no. 1 (May 20, 2022): 1–11, https://doi.org/10.1038/s41526-022-00201-y.

One of the big complaints from astronauts and cosmonauts onboard the International Space Station is the lack of quality coffee. Not just good tasting coffee, but also smelling the aroma, feeling the warm cup, and experiencing, ever so briefly, a moment of normalcy.

This issue was addressed with the delivery of a custom built espresso machine to the International Space Station. In 2015, Italian Espresso maker Lavazza, with the engineering support from Argotec and the Italian Space Agency, designed, built and delivered the ISSpresso.

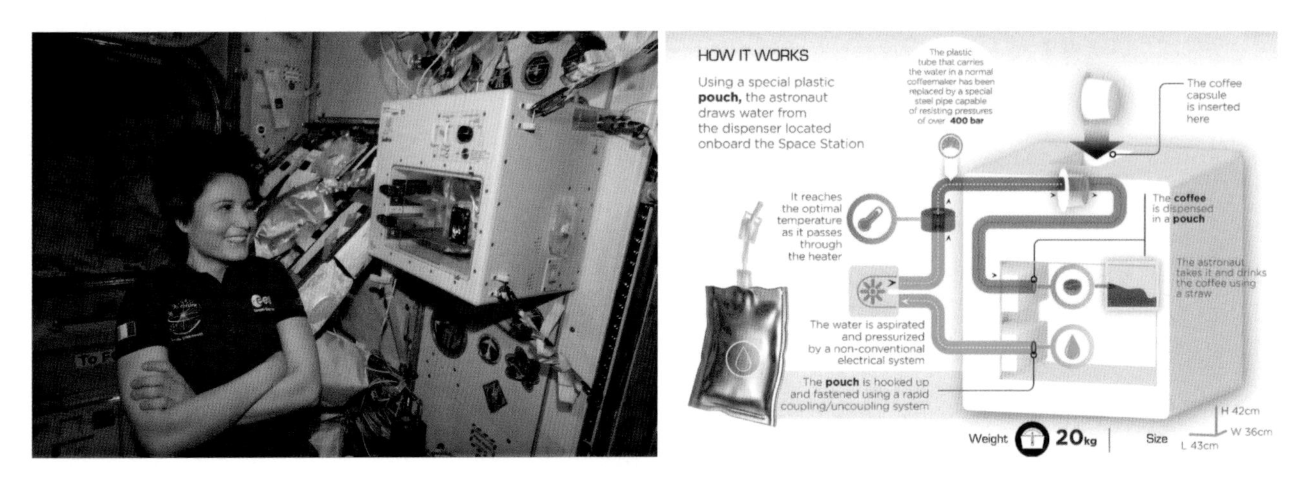

Left: astronaut Samantha Christoforetti next to the ISSpresso in the kitchen area of the ISS.[122] *Right: How the Isspresso works. Photos: NASA and Lavazza.*

Prior to the ISSpresso, the crew only had access to instant coffee. The instant coffee is in a drink pouch which is then filled with hot water to produce their coffee. The hot water is provided by the Potable Water Dispenser (PWD). The ISSpresso however, infuses coffee grounds with hot water under pressure to extract the espresso as would be done on Earth. The espresso is provided inside Keurig-style capsules, conveniently modified to prevent burst if exposed to vacuum, and properly packaged to avoid coffee grounds dispersion in case of accidental rupture.

ISSpresso does not only brew espresso, it can also produce hot beverages and consommé (such as chicken broth) for food hydration, depending on a wide selection of capsules. Additional options provide flexibility to the crew menu that is not currently available. For example, a dish of rice may have a variety of broth added to it for different tastes. When you boil water on Earth, the bubbles of steam stay evenly distributed in the water because of gravity. In space, these bubbles can come together and create pockets of really hot air.

It may not seem like much, but when you are traveling, being able to make a decent cup of coffee in precisely the way you like it has a huge psychological benefit.

---

122. "ISSEspresso | NASA Image and Video Library," NASA Image and Video Library, May 3, 2015, https://images.nasa.gov/details-iss043e160068.

Concept: Space Kitchen

Creature Comfort #7: Washing hands

The simple task of washing can be an adventure, if you are not prepared. As astronaut Chris Hadfield demonstrated in a video a few years back, when you wet a washcloth and then squeeze out the water, the water does not want to leave.[123] Surface tension causes the water to stick to both hands and the washcloth. Cleanup can be a mess. That is why they use baby wipes/wet wipes instead for most occasions.

Now that you've been given a crash course in the current state of affairs of food and drink in space, it is time for you to figure out how to make food preparation better. How do we make it less messy? What tools or appliances are needed to improve comfort and convenience?

The early Skylab space station had a dedicated kitchen for their astronauts. The International Space Station does not. It has a multipurpose area that can be adapted for communal gatherings and for cooking and eating. At other times the food-related items are put away so that the area can be used for other purposes.

For the next generation space station there should be a dedicated kitchen or galley, similar to ones onboard larger sea-going ships and commercial aircraft. The ultimate goal of course, is to automate and simplify the drink and food prep process to the point that a Star Trek-style Replicator is possible. My drinkbot projects were just for fun, but robot bartender companies like Bartesian, MAKR SHAKR, and SOMABAR exist today, and are making a profit.[124] Robot restaurants also exist today, such as Mezli in San Francisco and Spyce in Boston.[125] Plus several others are popping up in Asia.[126] So what would it take to create a semi- or fully-automated robotic galley in space?

---

123. *Wringing out Water on the ISS—for Science!* (International Space Station: Canadian Space Agency, 2013), https://www.youtube.com/watch?v=o8TssbmY-GM.

124. "Robotic Bartender Market Size to Grow by USD 678.8 Mn, Bars and Pubs Leveraging the Use of Advanced Technology Will Drive Growth—Technavio," PRNewswire, September 29, 2022, https://www.prnewswire.com/news-releases/robotic-bartender-market-size-to-grow-by-usd-678-8-mn-bars-and-pubs-leveraging-the-use-of-advanced-technology-will-drive-growth---technavio-301635702.html.

125. Lauren Saria, "This Restaurant Is Run Entirely By Robots," Eater SF, August 17, 2022, https://sf.eater.com/2022/8/17/23308389/mezli-robot-restaurant-open-menu-san-francisco.

126. Alina Turchenko, "Brand New Dining Experience: Top 5 Automated Restaurants," PaySpace Magazine, March 31, 2022, https://payspacemagazine.com/tech/brand-new-dining-experience-top-5-automated-restaurants/.

Reality Check: What if these fancy machines stop working? What if they break? What if the food in the vat spoils? What if you just want to make your own peanut butter and jelly sandwich without some darn robot? That is why kitchens will never go away, and people still want to DIY their own meals on occasion. So why not have both automation and manual food prep onboard a space station?

When I think of a kitchen, I think of the following elements:

- Refrigerator / freezer
- Stove / oven
- Sink
- Dishwasher
- Counter / food prep space
- Pantry
- Garbage / Recycling / Composting Bins

Now add some high tech additions to that collection:

- Foodbot (based on 3D printing tech combined with a bioreactor / fridge / oven).
- Meal or veg assembler (in a sealed chamber to reduce the mess).
- Drinkbot (mix and produce drinks with different flavors and different ingredients, also in a sealed chamber).

With these new tools, we begin to get a step closer to the Star Trek Replicator. One innovation at a time.

You may be asking: why can't you simply send the same appliances we have in a normal kitchen to the space station? The reasons should be obvious by now:

- Most appliances as designed are too heavy and bulky to fly up into space.
- Some components are gravity-dependent in order to work.
- Some components require convection currents of air (which is unavailable in microgravity).
- Some systems require natural gas or propane (which are obviously unavailable in space).
- The intense vibrations of a rocket launch tends to shake things apart.
- Most appliances are not as energy efficient as their space-bound counterparts would need to be.

When designing a kitchen for an earth-bound home, architects spend a lot of time on the ergonomics of the "work triangle." This is the convenience of the oven, sink and refrigerator in close proximity to each other. Also, the work height of the appliance is important due to the ergonomics of different size humans using these appliances. In space applications there is the concept of the "work envelope" which is the limited range of reach that space-suited astronauts have for grabbing things and moving around.[127] Fortunately, inside a space station the human is not wearing a space suit and should have a far larger work envelope for handling things.

So how do you design a kitchen galley that you can work in three dimensions? Where the work triangle is now a work sphere? Time for you to go back to the drawing board (or Computer Aided Design software, or LEGO blocks, or simply back to the kitchen for another snack).

From my personal experience, you find out quickly how many people can practically work simultaneously in an earthbound kitchen. My wife and I joke about "one butt" kitchens: where in theory two or more people can fit in the space, but they tend to get in the way of the person actually preparing dinner. I usually stay out of the way until the end of the dinner-making process and help with making drinks for the family.

Appliances: What would you do if you had to reinvent a kitchen appliance? One way is to start with the basics. What are your needs in the kitchen? What are the typical tasks? What can you live without? What do you absolutely need to have? How did people live without these things in the first place?

The following is a conceptual space kitchen design I have developed based on my years of research. It's not perfect, but my goal is to get you inspired about the possibilities of creating a real working kitchen in future space stations!

Three Dimensional Food Prep

Food prep on Earth can require a lot of coordination, though it may not feel like it. What if you were strapped to a pole upside down, and had to assemble a sandwich? Cutting up food items can be tricky. Imagine trying to slice an apple: the action of slicing will cause the reaction of your body to float in the opposite direction! Astronauts use a combination of Velcro, straps, and magnets to hold various utensils and items in place. Also, they use foot restraints to keep themselves from floating away.

Manual Food Prep: assuming the kitchen galley is cylindrical in nature, the food prep station should be in the middle of the kitchen. That way, you can have access to all the appliances in any direction. This is similar to a mobile counter in some of the more fancy kitchens on earth. This station is a flat round table connected via an arm to one wall (stowed away when not needed), consisting of elastic bands, Velcro strips, and magnets. These are used to hold down the various utensils and containers. Ideally this station also has a clear plexiglass cylinder or dome with armholes on it to limit the

127. Brand Norman Griffin et al., "Creating a Lunar EVA Work Envelope" (39th International Conference on Environmental Systems (ICES), Savannah, Georgia, USA: Society of Automotive Engineers, 2009), https://spacearchitect.org/pubs/SAE-2009-01-2569.pdf.

distance that items float around. For example, one piece of tortilla (or flatbread or Naan) could be floating around inside the food prep chamber while you are squeezing mayonnaise and mustard on a second piece you are holding. Lining the walls around the kitchen are the important appliances to help you make a meal: refrigerator, oven, drinkbot, foodbot, sink, pantry, and garbage/recycling. All of these devices should be within easy reach of the central food prep station. (Oh, and incorporating lots of handlebars and foot restraints to get around or stay stationary).

Robotic Food Prep ("foodbot"): With the advent of growing plant and meat cells in vats, we now have the ability to create food of any shape or configuration. Combine this with 3-D printing techniques, you can create a foodbot machine that can prep, cook, and inject a blob of food onto your platter. Some versions of this device require you to prep the food and then insert it into the machine for mixing, cooking, and dispensing. Other versions can be linked directly to the plant and meat growing vats. Either way, the final product, when done cooking, will sound a bell to inform the chef (that's you) the meal is ready to serve. Depending on the food, you could open the plexiglas door of the foodbot injection chamber, insert the covered plate or bowl installed with a special one-way valve, and then push a button to dispense the food. For example, if you were making an outer space hamburger, you would choose the type of meat or veg, then after it is done cooking, it will inject a heated burger onto your plate. Ideally, you would have a tortilla on the plate first. (Maybe someday someone will make a crumb-proof hamburger bun). Getting a drink would be similar: ask the drinkbot for a particular flavored beverage, hot or cold, and it will make it for you. It will fill up either a squeeze bag or specially-designed zero gravity drinking vessel inside a special chamber (once again, just in case messes happen). By the way, carbonated drinks do not work in microgravity. The CO2 immediately separates from the rest of the drink in microgravity.

I use the term "food" loosely, because in space it does not match the traditional chicken or hamburger or salad that we may be used to eating. Creating practical food in space requires a transformation of tastes and attitudes.

Cooking in Space

Most Earthbound ovens work via convection currents. As I mentioned above, there are no convection currents when there is no gravity. The way astronauts cook things is on a hot plate/toaster device that looks like a briefcase. The heat coils touch and warm up the plastic bag containing the food via conduction (All food is stored in some kind of vacuum sealed plastic bag). The "sous vide" cooking technique used by professional chefs is similar in concept: the food is cooked in a vacuum-sealed bag. Heat transfer occurs between objects via direct contact. So the heat from the coils warm the bag, the heated bag warms the food.[128] Unlike sous vide, the astronauts do not boil water to heat the bag of food, because of the messiness of dealing with the water (though this could be a potential invention to research).

128. Alex Delany, "So, What Is Sous Vide, Anyway?," *Bon Appétit: Cooking* (blog), January 24, 2018, https://www.bonappetit.com/story/what-is-sous-vide-cooking.

Microwave ovens are currently unusable because they use too much power, can leak microwave radiation and cause radio frequency (RF) interference.

Manual Cooking: Below are some other space oven concepts:

- **Space Oven:** The Zero G Kitchen Space Oven flew in 2018 which acts similar to a toaster oven, using heating elements and baked cookies.
- **"George Foreman Grill":** As a bachelor, one of my favorite ways of cooking burgers and fish was via this nifty double-sided electric heating grill. For a space-age version, I would enclose the grill in a box with a forced-air mechanism to simulate convection currents across the food. The air could be re-circulated to keep the smell inside. To collect the grease, there could be an absorbent paper towel-like material to capture it on the opposite end from the fan. Note: this could be a viable way to cook bacon safely in space, as long as you can keep the smell and the hot grease under control. The grill is also easily washable after use.
- **Water Boiler:** The ISSpressso can boil water easily. I would not be surprised if the venerable instant Ramen Noodles became available for astronauts, if packaged properly.

Robotic Cooking: Here is an example of the process with a foodbot:

1. Food material (plant or meat cells) is grown in a vessel. Also known as a bioreactor, larger ones are currently being built by the company Good Meat in Singapore and the USA.[129] So this is no longer science fiction.
2. The vessel is refrigerated to preserve the food as needed.
3. A human makes a meal request, either via voice commands or pushing some buttons. (It is good to have both options just in case one doesn't work).
4. The food vessel gets removed from the refrigerator and is partially heated. This heating would have to take place via some type of induction; in other words, something hot has to touch it to cause the heat to spread across the surface and eventually inside the vessel.
5. Using a combination of a peristaltic pump and/or a pressure fed system, a dollop of food is injected onto a heating container.
6. This container is transferred to a double-sided hot plate device using induction to heat the material. This would be inside something akin to a grill. There can be induction coils on top and bottom of the container to heat up the food. A couple of small fans help blow the heated air across the food.

---

129. Damian Carrington, "World's Largest Vats for Growing 'No-Kill' Meat to Be Built in US," *The Guardian*, May 25, 2022, https://www.theguardian.com/environment/2022/may/25/worlds-largest-vats-for-growing-no-kill-meat-to-be-built-in-us.

7. After a certain period of time, the food is transferred to a serving tray in a special chamber. This avoids food floating around.

8. There could be additional parts to the process, including putting the food onto a tortilla, Naan, sticky rice, or other non-flaky carbohydrate. Also, condiments could be dispensed automatically.

9. Please keep in mind, this is a conceptual exercise. Once you start working with engineers, the real thing gets more complicated quickly.

Keeping Cool

Refrigerators: Basically a refrigerator sucks heat from the inside and dumps it to the outside of the box. Take a look at your refrigerator at home. You may notice the radiator grill on the back. This is where the waste heat is removed from the fridge. Air circulation around it, known as convection, moves away the excess heat from the refrigerator. Because of gravity, convection currents work all the time. But in space and microgravity, there are no convection currents. So basically if you did place a conventional refrigerator in space, the heating coils on the back of the refrigerator would overheat, creating a hot bubble of air around the radiator of the refrigerator. This bubble would not dissipate. The next generation of space refrigerators are being developed today. They are using new techniques to manage energy and yet keep things cool.

Manual Refrigerator: There are two space fridges in orbit right now: SABL (Space Automated Bioproduct Lab)[130] and FRIDGE (Freezer Refrigerator Incubator Device for Galley and Experimentation).[131] Both were designed, built, and maintained by BioServe Space Technologies at the University of Colorado, Boulder. While SABL is exclusively used for science experiments, at least one of the FRIDGE units is going to be for food storage. Unfortunately, FRIDGE has the storage space the size of a microwave oven, so the astronauts have to choose carefully what to put in there.

For a next generation space station, there needs to be a much larger refrigerator, which is low maintenance, and can store a lot more foodstuff. A new prototype fridge is being developed by a team of researchers from Purdue University, Air Squared Inc. and Whirlpool Corporation. They are adapting the cooling technology of a household fridge to make what they claim to be the most energy-efficient model for space.[132] It is still small, but future versions may be larger.

---

130. "SABL Is BioServe's Next-Generation Smart Incubator," BioServe Space Technologies, University of Colorado Boulder, College of Engineering and Applied Science, October 22, 2018, https://www.colorado.edu/center/bioserve/spaceflight-hardware/sabl.

131. Jeff Zehnder, "New FRIDGE Could Bring Real Ice Cream to Space," Bioserve Space Technologies, University of Colorado Boulder, Ann and H.J. Smead Aerospace Engineering Sciences, College of Engineering and Applied Science, April 23, 2020, https://www.colorado.edu/aerospace/2020/04/23/new-fridge-could-bring-real-ice-cream-space.

132. Shi En Kim, "The Quest to Build a Functional, Energy-Efficient Refrigerator That Works in Space," Smithsonian Magazine, July 27, 2021, https://www.smithsonianmag.com/innovation/quest-to-build-functional-energy-efficient-refrigerator-that-works-in-space-180978281/.

Robotic Refrigerator: More than likely, a separate, dedicated refrigerator will be needed for the foodbots. It will be integrated into the larger system of food cooling, heating and prep as mentioned in the robotic cooking process mentioned above. The waste heat can be directed into the ECLSS.

Doing Dishes

Bowls, plates, pots, and pans are nearly useless in microgravity. Food floats away. The food itself needs to be sticky enough to stay on the plate or bowl surface, which most foods are not, and stuff is guaranteed to break off and clutter the area with food bits.

Onboard the ISS, nearly every food container is designed for single use. The "dishes and bowls" are actually hermetically sealed bags to keep things from floating around. Almost no food containers are open ended. Only the items that are guaranteed not to float away or have small particles in them are safe to consume with open lids. This system works, but it is very wasteful. There are tons of plastic and mylar container bits that are simply thrown away and cannot be reused. One of the main reasons is there is no practical way to clean them. Yes, they have towels, soap, and baby wipes to clean things, but they are not useful if the food container is a plastic bag or sealed straw. Also, as usual, water is a precious resource, so simply soaking them in a tub of soapy water will not happen.

For future space stations and on the homestead that you are designing, you need to have a goal of food storage containers and utensils that are reusable and easy to clean. They must also be able to keep any type of food item in place for storage, heating, cooling, or for actual consumption. This will be a major design challenge in the near future.

Oddly enough, you may be able to get inspiration via the baby-industrial complex. When our child was born, my wife and I were inundated with advertisements and recommendations for thousands of devices and contraptions for managing a baby: diapers, baby wipes, special water bottles, and so on. The baby drink bottles reminded me of some of the NASA drink containers, since they are designed to not spill in any direction. Here are some example baby products that could be useful in space:

- Reusable baby food pouches with a resealable bottom, which could be cleaned out and reused.
- Heatable and reusable plastic containers.

Only the squeeze bags would work in space though, due to the liquids sticking to the surfaces. By the way, did you know the manufacturers of "Capri-Sun" kids drinks actually supply NASA with larger versions of those squeeze bags?[133]

---

133. "Our History | Learn More about the Capri-Sun Story | Capri-Sun UK," *Capri Sun Great Britain* (blog), accessed January 14, 2023, https://www.capri-sun.com/uk/about-us/history/.

Here is another inspiration source: doing dishes while camping. Basically, it is really tedious. Here is how I have cleaned dishes while desert camping, using the least amount of water:

1. Get a wash cloth and add a couple drops of soap into a corner.
2. Pour a little water into that same corner.
3. Rub the soapy and wet part of the washcloth onto all dirty parts of the dish or utensil.
4. Use your fingers to dislodge the more difficult food bits.
5. Rinse off the dish and the cloth corner with water.
6. Dry the dish or utensil immediately with the non-wet parts of the washcloth.
7. Pack it away.

This process can be made even simpler by using wet towelettes (baby wipes).

How would you clean dishes in space? Most likely the only things you could clean are utensils (forks, spoons, knives, chopsticks) and reusable plastic containers. Here are some concepts to get you started:

Manual Cleaning: The Space Sink is designed to emulate an Earthbound kitchen sink, but with everything sealed up to prevent water and debris from floating everywhere. It consists of a cylindrical chamber with a plexiglass dome on top. The whole chamber is coated with a hydrophobic and oleophobic coating. (Hydrophobic means “resists water” and oleophobic means “resists oils.”) There are two glove box openings for you to put your hands into the chamber. The rubber gloves are also covered with a hydrophobic/oleophobic coating. There are four hoses/tubes inside: one for air, one for water, one for soap, and one for vacuum. The vacuum tube has a mesh screen to prevent objects (like small utensils) from being sucked in. The water temperature and pressure can be controlled by controls on the outside. To insert dishes or utensils, there is a small double-door airlock in front. Here is a rough idea of the cleaning process:

1. Open the outer airlock door and insert the dirty food containers and utensils into the airlock.
2. Close the outer airlock door.
3. Put your hands into the rubber gloves in the glove box.
4. Activate the air, water, soap and vacuum.
5. Open the inner airlock door and grab an item. (Note: the vacuum may pull the items into the sink chamber)
6. Grab the washcloth (which is hooked to the inside of the sink)
7. Inject some soap into a corner of the cloth

8. Inject some water into the same corner
9. Rub the cloth to build up a lather. (no worries about water or soap floating around due to the vacuum).
10. Rub the soapy and wet part of the cloth onto the dirty part of the dishes.
11. Use your fingers to dislodge the more difficult food bits. (There are other tools, like brushes, that could assist in this process.)
12. Rinse off the item. Since you are going to have a water blob surrounding everything, this will require being near the vacuum.
13. Grab the vacuum hose and suck up excess water, soap, and debris.
14. Use the dry part of the cloth to dry off the rest of the item, if there is any water left on the item.
15. Use the air blower to blow off any last bits of water.
16. Put the item back into the airlock (assuming velcro or magnets are attached to it).
17. Squeeze out the washcloth. Since the water will stay glommed to the washcloth, do this in front of the air blower and near the vacuum. Dry it out as much as possible.
18. Put the washcloth back on its hook, or put it in the airlock for cleaning. (See "Laundry that Floats" in a later chapter).

As you noticed, the process is very similar to the Earthbound camping process I just mentioned above.

Where does the water, soap, and food debris go? To reuse the water a few more times during washing, the vacuum pump pushes the water through a filter. The waste water passes next to a light bulb which blasts ultraviolet (UV-C) rays into the used water, killing most pathogens. This semi-sterilized water is redirected back into the sink hose for reuse. After the final rinse, the waste water gets vented into the space station's life support system for processing, and eventually back into the space garden. Plants plus bacteria can assist with breaking down the food bits and also help the garden grow. Also, there could be a composter device, which uses a combination of high temperatures, moisture and bacteria to break down the food into soil. In the future, it may be possible to use soil as well as hydroponics to grow plants. More about waste management later in the chapter.

What if you have a lot of items to wash and clean? That is where the washbot comes in.

Robotic cleaning ("washbot"): A washbot handles large quantities of utensils and food containers at once. Using some of the concepts from the space sink (including the hydrophobic/oleophobic coating), each item needs to be held down by clamps onto a special tray.

1. Once each item is locked down, the tray is inserted into a larger airlock designed for bulk items.
2. When the outer airlock is sealed, the inner airlock door opens and the tray moves into the washing chamber.
3. The washing chamber has a series of air and water jets + soap to blast each item with soapy water.
4. The machine uses brushes and vibrations to knock the food particles off.
5. Since everything will be coated in one giant blob of water, a series of air jets + vibrations will start disrupting the surface tension. At the same time, the vacuum hose will turn on, sucking water, soap and debris, while the air jets and vibration loosen up the water.
6. A heater turns on to help with water evaporation. The vacuum is still sucking away at the mist and particles.
7. After a predetermined time period, the system stops, and the dish tray is moved back into the airlock.

As per the Space Sink, the wastewater is filtered, blasted with UV-C rays, and then pumped back into the washbot. When the washbot is done, the final round of water, food, and soap go through the life support system, the composter, and eventually the space garden. The soap has to be designed to break down into a plant-friendly chemical.

Waste Disposal and Recycling

On earth, most of the garbage goes to a dump. Some of it is recycled. In the end, it becomes a big mess. In space, astronauts and cosmonauts do basically the same thing: eject garbage out the airlock, or fill up a cargo shuttle and direct it to burn up in Earth's atmosphere. With the issue of space garbage in Earth orbit becoming more prominent every day, this is not a very good option. Remember, these objects are flying at thousands of miles an hour. They could one day become projectiles that may damage satellites and spacecraft. Could there be a way to reuse trash? Or could you plan ahead and design items to break down more easily?

In 2002, NASA developed a prototype composting device called the Space Operations Bio Converter (SOB).[134] It was a composter consisting of a rotating drum that contained waste for decomposition. It's nothing like your backyard compost heap. Its goal was to find an efficient way to break down waste while extracting as many nutrients as possible. They tested different bulking materials

134. Abner A. Rodriguez-Carias et al., "In-Vessel Composting of Simulated Long-Term Missions Space-Related Solid Wastes" (NASA John F. Kennedy Space Center, Publication: 2002 Research Reports: NASA/ASEE Fellowship Program, December 1, 2002), https://ntrs.nasa.gov/citations/20030062829.

and temperatures to improve the biodegradability. The experiment showed potential, but was never followed through. In the study, they divided the garbage into two categories:

- Fast biodegradability: bread, meat, vegetables, any plant matter
- Slow biodegradability: food packaging, paper, tape, plastic

Not surprisingly, plastic has the slowest biodegradability, and the food bits the fastest.

Manual Waste Management: You could do what everybody else does, and simply eject waste, packaging, and bad food out the airlock. If the space station doesn't maneuver much, that garbage will just simply follow alongside the space station. If the space station has a propulsion system to cause it to maneuver or change orbit, then the waste will be left behind. Sooner or later though, that garbage is going to have to be dealt with.

Robotic Waste Management: There are several automated or semi-automated systems which could help process the waste:

- **Pyrolysis:** Waste packaging and food could be put into a device that would heat it up over 500ºC (932ºF) without oxygen, providing enough heat to deconstruct many food packaging as well as food items. The pyrolysis of biomass produces three products: one liquid (bio-oil), one solid (bio-char), and one gaseous (syngas).[135] Technically, the biochar could help with the gardens and absorb CO2. The Bio-oil and syngas could be used as fuels similar to gasoline (that is a whole other area to research). Thermal Depolymerization is a variation of pyrolysis that can break down plastics and most organic matter to a crude oil that could be refined into fuel.
- **Incineration:** In contrast to pyrolysis, Incineration uses oxygen to reduce the waste to ash. It requires a second step to reduce the waste further to ash.[136]
- **Anaerobic digestion:** A safer way to dispose of waste, Anaerobic digestion uses enzymes and bacteria to ferment or digest organic materials without oxygen present.[137] Basically, helping nature do what it normally does, but a bit faster. Enzymatic depolymerization is another less energy intensive technique that could break down plastics.

All of the above options require lots of energy and room to process waste. There is still going to be some kind of leftover material, like ash or certain types of gas, which may need to be ejected from

135. "What Is Pyrolysis?," U.S. Department of Agriculture, Agricultural Research Service, September 10, 2021, https://www.ars.usda.gov/northeast-area/wyndmoor-pa/eastern-regional-research-center/docs/biomass-pyrolysis-research-1/what-is-pyrolysis/.

136. National Research Council (US) Committee on Health Effects of Waste Incineration, *Waste Incineration & Public Health* (Washington (DC): National Academies Press (US), 2000), http://www.ncbi.nlm.nih.gov/books/NBK233629/.

137. "Anaerobic Digestion Explained," *RENERGON—ANAEROBIC DIGESTION* (blog), July 25, 2021, https://www.renergon-biogas.com/en/anaerobic-digestion-explained/.

the space station, or they could be used in the life support system, or used as a power source, or used as fertilizer in the space garden. There has not been research as to how these systems would work in microgravity. Your design challenge is to figure out how to recover as many bits of material for future use.

Other options for waste management which have not been explored is radiation and the hard vacuum of space. We know that many materials tend to break down over time if left exposed to direct sunlight. When you go outside the protective atmosphere of Earth, there are even more destructive elements like UV-C, gamma rays, and cosmic rays. How can you use these as tools in designing a waste management system for space?

Speaking of waste, let's move onto the next chapter which discusses the number one and number two things that all people worry about when they go traveling: how do we go to the bathroom and what do we do with the waste products?

Chapter 3:

When Nature Calls

Creature Comfort #8: Toilets/Bidets

In a nutshell, toilets suck. Literally and metaphorically. From a design and usability standpoint, major improvements are needed to improve the quality of the toilet experience.

How do you go to the bathroom in space? The answer is: differently. Stuff floats. Stuff follows you. How do you prevent that stuff from doing that? How do you keep things clean? How do you manage the smells? How can you keep the process as simple as possible? What do you do with waste products? All these factors need consideration when designing a new toilet and the necessary wastewater management system for future space stations. This chapter will discuss the current state of affairs with NASA and Russian Space toilets and offer some suggestions for improvement. How would you design a nice looking toilet, AND a system-wide processing system for #1 and #2?

If you are wondering why I am currently obsessed with this topic, consider this: most of us are spoiled. You are probably used to a certain way of going to the bathroom, and have never even thought that there are other ways of "doing what needs to be done." When you go camping, you are probably thinking about getting closer to nature. Guess what? You are gonna have to learn how the animals pee and poo in the woods real quick, unless you have a restroom or outhouse nearby.

Going to the Bathroom on Earth and in Space

Here is a comparison of different levels of doing #1 and #2 on Earth, from the most "civilized" to "roughing it." You will discover that in space, we are only a little bit beyond the "roughing it" stage.

Level 1 Most Civilized: When you pee or poo, everything goes into the venerable toilet, then you wipe your bum and then you pull down the flush handle. Most Earth toilets use gravity and water pressure to move urine and feces from the toilet bowl, through the sewer system, and eventually to a central wastewater treatment plant. The treatment plant processes the wastewater in a series of filtration systems. When the final water is tested to be clear of most pathogens, it is then dumped in the local river. This is common for most cities and the suburbs in most Western countries.

Level 2 Homesteading It: For folks living in farms and remote areas, septic systems are more common. The peeing and pooing part of the process is basically the same, but the home may be using a septic system. With a septic system, wastewater is directed into a septic tank, where the solids slowly separate from the liquids, and eventually the waste liquids go into a drainfield.[138] The drainfield could be indistinguishable from a field or garden, but you never, ever want to play or grow things in it. The soil and the plants act as the wastewater treatment plant: mother nature does the hard work of dealing with E-Coli and other pathogens. Depending upon the environmental and climate situation, it can take up to a year for the pathogens to be eliminated.[139] The solids will either break down and be scattered into the drainfield, or not, and will have to either get pumped out annually from the septic tank and sent to a treatment plant for processing, or they get sent to the dump. You could also dry the feces to make it into compost for fertilizer...after a year of drying out.[140]

Level 3 Road Warrior: If you own an RV or camper/caravan, you may be familiar with greywater and black water. There are two separate tanks in the camper: one for sink or shower water (greywater), and the other for the toilet (black water). These tanks have enough capacity for a family of four to enjoy the weekend "boondocking": camping off the grid, unconnected to sewer or power, far from civilization. Sooner or later though, the tanks will get full and it's time to go to the nearest dump station (a polite name for a designated underground sewage tank).

Level 4 Roughing It: If you are hiking in the great outdoors with only a backpack and a small tent then it is time to really get back to nature. After you set up camp (preferably on higher ground), you may find some bushes nearby to do your duty. Pro Tip: Make sure you aim downhill, or gravity will tell you are

---

138. Lloyd Kahn, *THE SEPTIC SYSTEM OWNER'S MANUAL*, 2nd Edition (Bolinas, CA: Shelter Publications, 2007), https://www.shelterpub.com/building/the-septic-system-owners-manual.

139. Gord Baird and Ann Baird, *Essential Composting Toilets: A Guide to Options, Design, Installation, and Use*, 1st Printing (British Columbia, Canada: New Society Publishers, 2018), https://newsociety.com/books/e/essential-composting-toilets.

140. Joseph Jenkins, *The Humanure Handbook: Shit in a Nutshell*, 4th ed. (Grove City, PA: Joseph Jenkins, Inc., 2019), https://slateroofwarehouse.com/Books/Joseph_Jenkins_Books/Humanure_Handbook.

doing it wrong. As for #2, you can either get one of those camping toilet bags, or go hardcore and take a squat over a makeshift hole in the ground. Hope you brought some toilet paper and a small shovel!

In space: The toilets on spacecraft consist of two components: a hose for urine collection, and a cylindrical tube for collecting feces. A vacuum helps suck the urine into the tube and into the life support system for processing. In current spacecraft, the waste typically gets ejected out a tiny airlock. (It is fun to watch the urine instantly turn into icicles!) Onboard the ISS, NASA has recently improved upon its urine processing system, where most of it can be chemically processed back into water. Feces are sealed into bags and thrown into the next cargo ship to burn up in the atmosphere. The cramped conditions onboard spacecraft and the ISS mean that space toilets have to be as small as possible. Another difficulty is that the general taboo around human excrement means that not enough research has been done exploring true recycling of human waste. I will cover this topic later in this chapter.

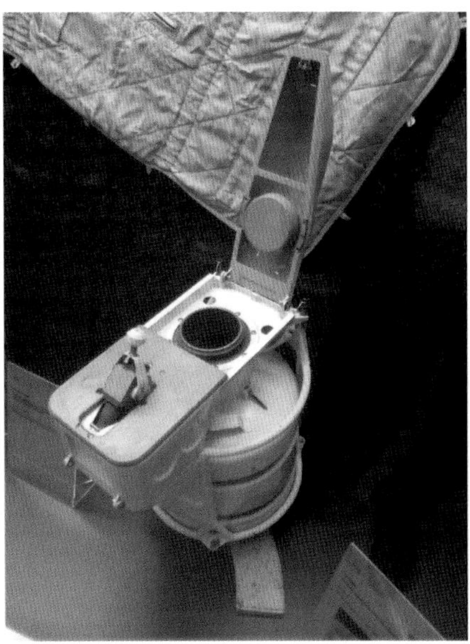
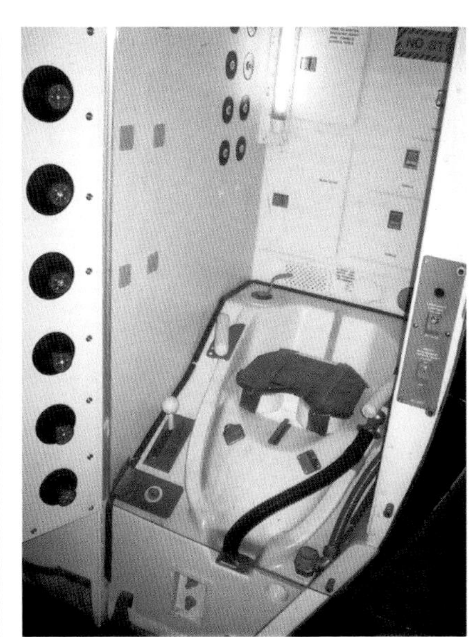

Left to Right: Toilet in Zvezda Service Module on the ISS, Russian MIR Space Station Toilet, Space Shuttle Toilet. (Can we design these things to look nicer?) Source: Wikipedia.

The Struggle is Real: Reinventing the Poop System

As you can see, the current space toilet systems are halfway between the Earthbound "Road Warrior" and "Roughing It" analogies than to modern home systems. With microgravity, everything floats, so this is why it is difficult to create a good toilet system. NASA and the Russian space agency have struggled with this most human of activities since the beginning of the Space Age. Even SpaceX, the current rockstar space company, had challenges with a failing toilet during the

2021 flight of the Inspiration4 crew. A tube from the capsule's toilet that funnels waste into an internal tank broke loose and leaked fluids into a fan, which sent urine throughout an area beneath the capsule's interior floor. Once discovered, SpaceX's engineering team did some research on the possible corrosion of the spaceship's aluminum hull from the floating urine.[141] They determined that the urine evaporated long before damage could occur. So the mission continued. In the meantime, the Inspiration4 crew had to resort to a classic stand by: space diapers![142] It was standard operating procedure for astronauts to wear a diaper under their spacesuit, just in case. "Just in case" happened, and they were ready.

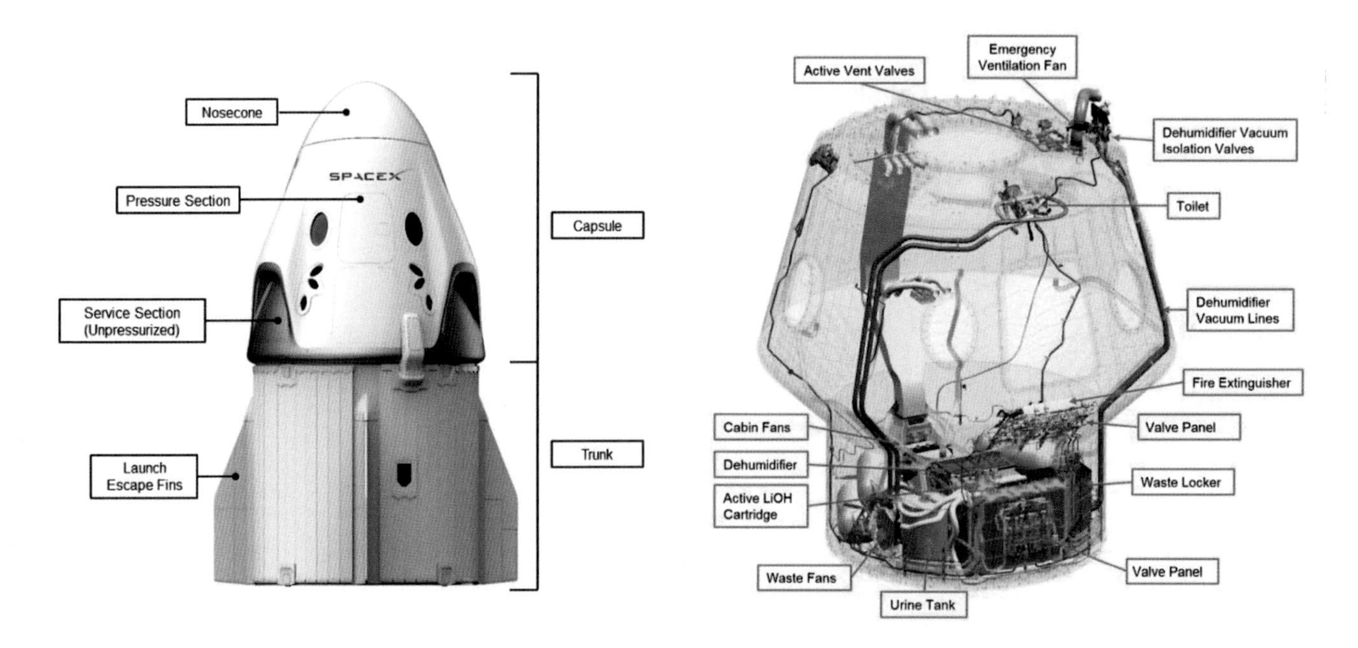

The Crew Dragon spacecraft (left), and its Environmental Closed Life Support System (ECLSS, right). Note the location of the toilet up near the top airlock. Source: Jason Silverman, et.al. "Development of the Crew Dragon ECLSS," Page 3, ICES-2020-333, 2020 International Conference on Environmental Systems.[143]

141. Joey Roulette, "SpaceX's Toilet Is Working Fine, Thanks for Asking.," *The New York Times*, November 11, 2021, sec. Science, https://www.nytimes.com/2021/11/10/science/spacex-toilet-diapers.html.

142. Joey Roulette, "The Toilet on the Crew Dragon Capsule Was out of Service. The Crew Had to Use Diapers.," *The New York Times*, November 9, 2021, sec. Science, https://www.nytimes.com/2021/11/08/science/spacex-diapers-toilet.html.

143. Jason Silverman, Andrew Irby, and Theodore Agerton, "Development of the Crew Dragon ECLSS," vol. ICES-2020-333 (2020 International Conference on Environmental Systems, 2020 International Conference on Environmental Systems, 2020), 11 pages, https://ttu-ir.tdl.org/bitstream/handle/2346/86364/ICES-2020-333.pdf.

In October 2020, NASA launched a new, "universal" space toilet to the International Space Station. Called the Universal Waste Management System (UWMS), it is lighter than its predecessors and is adaptable for the ISS and future long range space missions to the Moon and Mars. [144]

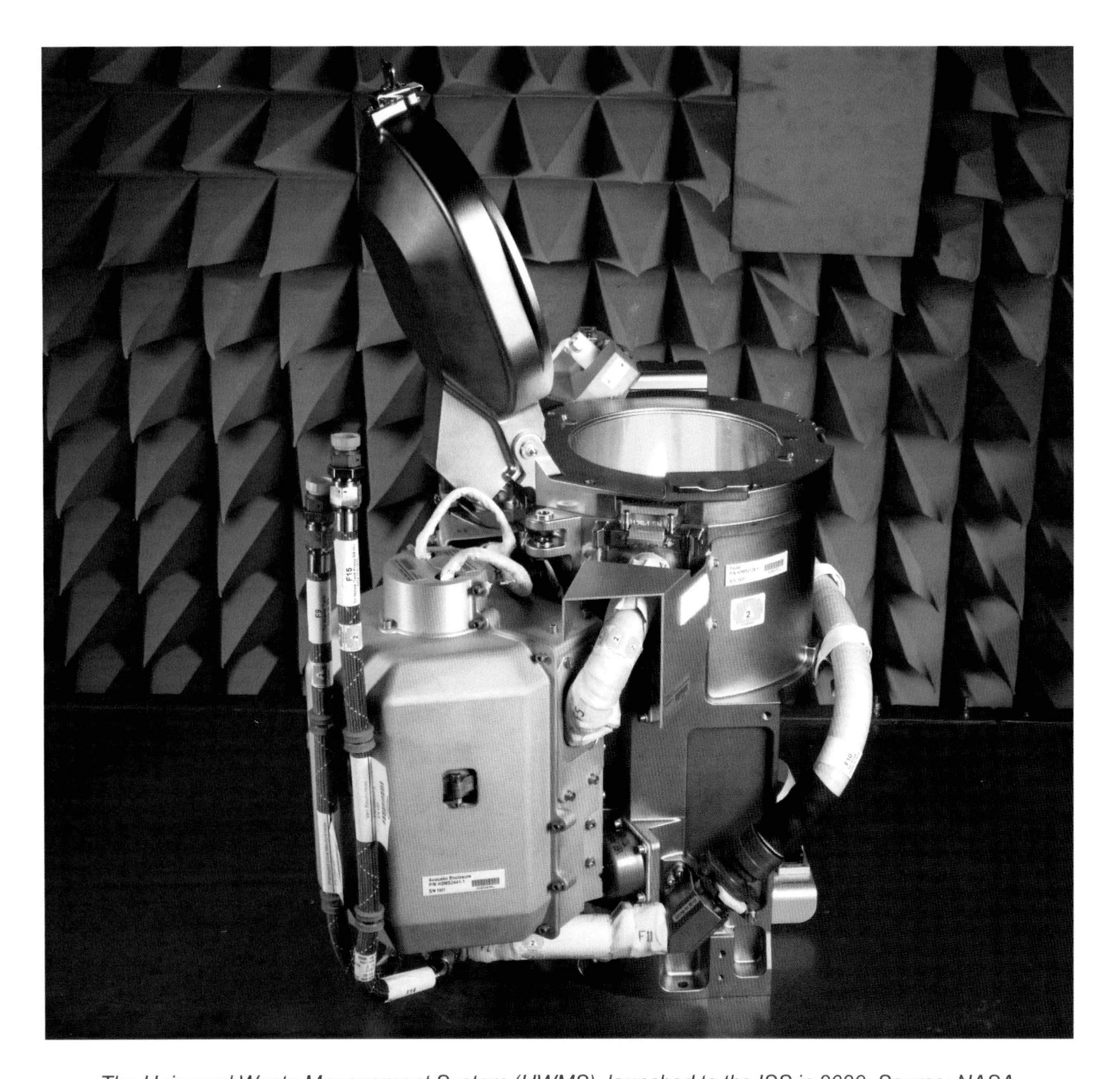

The Universal Waste Management System (UWMS), launched to the ISS in 2020. Source: NASA.

144. Thomas J. Stapleton et al., "Development of a Universal Waste Management System," in *43rd International Conference on Environmental Systems* (43rd International Conference on Environmental Systems, Vail, CO: American Institute of Aeronautics and Astronautics, 2013), https://doi.org/10.2514/6.2013-3400.

The UWMS is designed to chemically pre-treat urine in a controlled manner. This reduces the amount of ammonia gas that is generated as urine breaks down.

The urine funnel was redesigned with the female anatomy in mind (it only took a few decades!).

The bathroom stall now has two compartments (which means two toilets for large crews).[145]

This is all well and good, but *can they design it to look better?* NASA's focus is engineering a toilet that works better in microgravity, and design aesthetics are not a priority. Also there is very little room in the ISS for flourishes, since sometimes the astronauts have to put on their plumber's hats and repair the things. Still, would you want to poop into a metal tube? Foot restraints become critical here, since your body tends to float away. A private astronaut told me a rather graphic story about going to the bathroom: "that poo floats, and sometimes it floats towards you!" Thus the need for lots of baby wipes and paper towels.

Did I mention that baby wipes are standard equipment onboard the ISS? I use them extensively for camping and for general house cleanup. When you raise a child, the power of baby wipes becomes apparent, and soon you want to use them for cleaning up all kinds of messes around the house!

As for peeing, you could pee upside down and it would not matter. There is no up or down in microgravity. Hopefully, suction from the vacuum hose is strong enough…

The US astronauts have a hilarious term for peeing and pooing at the same time: "dual ops". So military! I can almost hear it now on the comms: "This is Mission Control. Astronaut 'so-and-so' has successfully completed Dual Ops without any collusion." (Hey, I gotta put a Dad Joke somewhere in this book!)

The biggest challenge of toilets in space is the obvious one: how to capture the feces? An aerospace expert who worked on the UWMS explained that getting the feces away from the butt and into a container (like a ziplock bag), can be one of the most humiliating activities that an astronaut has to do. It was bad enough that a test model of the US Space Shuttle toilet had a camera system inside the toilet pointed at your anus to help you assure "perfect alignment".[146]

What do you do with waste products? NASA has worked really hard to find ways to recycle urine back into water. But are there ways to reuse the feces?

Technically urine could be considered greywater because it can be processed and recycled. But black water, feces, toxic chemicals and other stuff like that are not easily processed. ISS toilets currently are designed to hold up to 30 bags of feces. They eventually get put into a cargo shuttle which is eventually burned in the atmosphere. But what if we can take the potential poop and use it for something else?

145. Melissa McKinley et al., "NASA Universal Waste Management System and Toilet Integration Hardware Delivery and Planned Operation on ISS," in *50th International Conference on Environmental Systems*, vol. ICES-2021-403, 2021, https://ttu-ir.tdl.org/bitstream/handle/2346/87295/ICES-2021-403.pdf?sequence=1&isAllowed=n.

146. Organic Marble, "Was There Really a Shuttle Toilet Training Device with a 'Boresight Camera'?," Forum post, *Space Exploration Stack Exchange*, June 25, 2020, https://space.stackexchange.com/q/40338.

Closed Ecological Systems

For thousands of years, manure of all kinds, human or animal, when dried, was used as a type of fertilizer. For some cultures such as China it was part of the natural flow of society: the "nightsoil" was collected by farmers in the towns and cities, and then they carted or boated it back to the rural farmland where it became fertilizer. In Western cultures, people simply dumped their waste in the streets. Diseases such as typhoid and cholera ran rampant. As cities became larger, it became obvious something needed to be done. Sewers were developed, but instead of routing the waste to farms, it went into rivers, lakes and oceans. Huge amounts of freshwater was used to move that waste from point A to B, which is a key factor to the water crisis that the human race is just starting to face in the 21st century.[147]

So what does this history lesson have to do with space stations? The problem is the viruses and bacteria that thrive in that environment. Using our homestead analogy, we must find a way to process greywater and black water to be safe for farming and other things. A disease epidemic onboard a space station could mean disaster!

Concept: Bidet in Space!

While I was researching space toilets, one of my friends on Facebook jokingly asked "what about bidets in space?" I had to laugh as well. Bidets are those fancy toilet accessories which reduce paper usage by giving a little "squirt in your bum," as the Brits would say.[148] While popular in many parts of the world, off-world versions would need a major redesign due to surface tension issues (water blobs sticking everywhere). One bidet company, Tushy, jokingly offered to build a space bidet for SPACEX after the Inspiration 4 mission reported having problems with their space toilet in 2021.[149] Silliness aside, I later realized that maybe, possibly, my friend was onto something.

Although we cannot take an existing toilet on Earth and bring it into space, you can at least create a design aesthetic that reminds you of a home toilet. That unique oval shape, that soft cushion seat. (That may actually be it, come to think of it). Everything else has to accommodate for the bizarre behavior of weightlessness.

First and foremost you need foot straps. Since your body tends to float away, there also needs to be a belt to keep your body close to the toilet seat. There should be handles on the side walls and ceiling of the bathroom chamber. This will allow a person to hold onto something while going to the bathroom. I developed these ideas after I watched a hilarious conversation of astronauts talking about the space shuttle toilet training.[150]

---

147. Mark Nelson, *The Wastewater Gardener: Preserving the Planet One Flush at a Time*, 1st Edition (Santa Fe, NM: Synergetic Press, 2014), https://synergeticpress.com/catalog/the-wastewater-gardener/.

148. Nafeesah Allen and Lowe Saddler, "What Is A Bidet? How Does It Work?," Forbes Home, November 1, 2022, https://www.forbes.com/home-improvement/bathroom/what-is-a-bidet/.

149. VICTOR TANGERMANN, "As Spacecraft Toilet Rumors Swirl, Bidet Company Pitches Elon Musk," Futurism, September 23, 2021, https://futurism.com/bidet-company-elon-musk.

150. *Shuttle's Toilet Requires Special Training*, vol. 1, STS-132: Behind the Scenes (NASA YouTube Channel, 2010), https://www.youtube.com/watch?v=m1wwzwvfsC0.

In the design concept I am proposing, the toilet itself is shaped similar to a horse saddle, so that your legs are spread slightly apart. It has two parts, both coated with the same hydrophobic material as the space sink in the kitchen and in the space shower tube (which I discuss in the next chapter).

The pee funnel is designed for both the male and female anatomy. After peeing into the funnel the user closes the funnel cap and presses the flush button. A small amount of water is injected into the funnel tube to glom onto the urine, then a vacuum is activated to suck the water and urine. The urine and water is directed to the urine processing system to convert into water. (Also, you may want to redirect some of the ammonia from the urine to the space garden as a fertilizer.)

The toilet seat is designed to seal around the butt and use a blob of water to capture the feces. I am thinking the seat opening should be a "teardrop" shape (similar to Donald Petit's original zero gravity coffee cup design) to help naturally guide the water. Yes, this is basically a "space bidet" as the water is connecting your anus to the toilet seat with water. The idea is to use the surface tension of the water to surround and capture the feces as it is coming out of your body. When you are done defecating, you can press the button to activate the vacuum. If all goes well, the vacuum will suck the water and the feces it into a collection chamber. If things tend to stick, maybe add a vibration device to shake the vacuum tube and the bidet seat to separate your bum from the water and feces. This idea of creating a hermetic seal between one's butt and the toilet will require a lot of testing of different seat shapes and water management. It will take some trial and error to figure out a good system. A lot of stinky, icky trial and error. It's a dirty job, but somebody's gotta do it.

Processing the Pee and Poo

The toilet's goal of keeping the urine and feces separate is to make it easier to process them. For the urine, we can use NASA's urine processing system to convert it back into water. For the feces, there must be some kind of system that could break it down into fertilizer for the space garden. The system could be some kind of super composting process. One idea is to use the vacuum of space and the natural space radiation to help kill pathogens. This could be as simple as opening a special airlock door in the toilet exhaust chamber that keeps the feces in, but exposes the material to space. By coincidence, humans have been using radiation to kill germs on Earth, in the form of UV-C.

The Power of UV-C

For extreme camping, especially folks who like to do endurance hiking for weeks in the wilderness, there are devices called Steripens.[151] Instead of having to lug freshwater with you, you can collect water from the streams, filter it out as much as you can, and then use the Steripen to blast UV-C light

151. Lisa F. Timmermann et al., "Drinking Water Treatment with Ultraviolet Light for Travelers—Evaluation of a Mobile Lightweight System," *Travel Medicine and Infectious Disease* 13, no. 6 (November 1, 2015): 466–74, https://doi.org/10.1016/j.tmaid.2015.10.005.

into the bottle. After about 90 seconds or so, most pathogens are killed. This makes the mountain water (more or less) safe to drink.

There are three types of ultraviolet radiation: UV-A, UV-B, and UV-C.[152] UV-A is the good stuff: it gives us suntans, helps with production of Vitamin D in our bodies, and helps build bones. UV-B is not so good: it causes sunburns, and with long enough exposure, skin cancer. Most UV-B is blocked by the ozone layer of planet Earth's atmosphere, so one has to live in sunny areas for years to get the ill effects. UV-C is another story. It is completely absorbed by the ozone layer and atmosphere. That means plants and animals have never learned to adapt to it, so it can be extremely dangerous when exposed to it. What makes UV-C so dangerous is that it literally rips apart DNA molecules, making them unable to replicate. It is so effective, that UV-C robots are now mobilized in some hospitals to disinfect a room after patients and personnel leave it.[153] There are UV-C lights in some HVAC systems to filter the air.[154] Direct exposure to UV-C by a human is also dangerous, because of the above mentioned DNA damage that happens. There are limits: the lights only work on things that are in direct contact, close range, or in non-cloudy water. If your body and face are covered by clothing, or an item is in shadow, the effects are much weaker. During the COVID-19 pandemic, some local hardware stores sold UV-C sterilizers for cell phones and other small items. Just don't expose the phone for too long. UV light is known for breaking down plastics and other materials as well.[155]

So in the case of urine and feces processing in space, there needs to be a pre-filter to clarify the liquids before blasting it with UV-C. More than likely NASA has something like this integrated into its space station life support system.

Side story: I took a tour of the local sewage treatment plant in my community. I learned that they have a seven step process for filtering out debris and killing various pathogens. All that was left at the end of the process is a nutrient-rich compost that is trucked to a local farm as fertilizer. The remaining liquid waste is filtered down to the microbial level, and then tested with some small fish. If the fish stay alive for five days in that water, then they would dump the processed water into the river.[156] I was told that to process it back to potable and safe drinking water, it would cost millions of dollars in new equipment and systems to finish the job for a city's worth of wastewater. (In my opinion, this could save on billions in future water usage fees and prevent future droughts, but I am not an expert).

For our future space station, we do not have to worry about processing millions of gallons of wastewater. There should be a way to efficiently convert a modest amount of something that is normally icky into something that can be useful for sustaining life onboard.

---

152. "UV Radiation," Centers for Disease Control and Prevention, July 5, 2022, https://www.cdc.gov/nceh/features/uv-radiation-safety/index.html.

153. Evan Ackerman, "Autonomous Robots Are Helping Kill Coronavirus in Hospitals," IEEE Spectrum, March 11, 2020, https://spectrum.ieee.org/autonomous-robots-are-helping-kill-coronavirus-in-hospitals.

154. "New Type of Ultraviolet Light Makes Indoor Air as Safe as Outdoors," Columbia University Irving Medical Center, March 17, 2022, https://www.cuimc.columbia.edu/news/new-type-ultraviolet-light-makes-indoor-air-safe-outdoors.

155. "UV and Its Effect on Plastics: An Overview," Essentra Components, January 23, 2019, https://www.essentracomponents.com/en-us/news/manufacturing/injection-molding/uv-and-its-effect-on-plastics-an-overview.

156. "Vallejo Wastewater, CA | Official Website," accessed May 11, 2023, https://www.vallejowastewater.org/.

Chapter 4:

Showers

Creature Comfort #9: Hot Showers

Nothing beats a warm shower to relax the body and mind. Unfortunately, it is very difficult to do in a microgravity environment. Why? Because water likes to stick to things, and is very hard to clean up.

Creature Comfort #10: Fluffy Towels

Towels of any kind are nice to have. Fluffy or not, the important thing for towels is to absorb liquids. Wrapping your body with a warm fluffy towel after a shower is nice, if you can keep it from floating away.

When I first experimented with extreme camping, out at the Black Rock Desert in Nevada, it was a shock. For the first time, I had to pack everything: not just toiletries, but food, water, and toilet paper. Camp showers were a rickety affair: usually just a bag of water heating up in the sun. I washed one body part at a time, soaped that part down, rinsed off, and moved on to the next body part. Ideally you did this activity over an "evaporation pond," which is typically a giant black plastic sheet anchored to the ground, so that the wind and sun can safely evaporate the greywater without contaminating the "playa," the dry lake bed in which Burning Man takes place. It was a messy, incomplete shower, but it still felt good when you did it in the 38°C (100°F) weather, preferably in between dust storms.

On most days though, I improvised. Sometimes I got the corner of a towel wet, added soap, and focused on cleaning my "pits and bits." In later years, I started to rely on the miracle of childcare: baby wipes. Baby wipes are designed to clean up pee, poo, vomit, and other messes that a baby or small child makes. I will tell you, it is a simple and amazing tool for cleaning the human body. You and your child could be in the car, at a restaurant, a theme park, or any random place, and when you or the child makes a mess, the baby wipes can clean up anything.

When I upgraded my Desert Camping routine to an RV camper 15 years later, having a little shower inside was like I moved to the First Class cabin in an airline! It was sheer luxury to take a real shower in a harsh environment! You still have to ration your water usage, but the act of getting a full body wash after a couple days getting "playafied" (covered head to toe in dust) felt like Heaven!

The psychology of taking a shower is a big deal. It is not just about getting the body clean. It is about relaxation, and recharging your "mental batteries."

The showers onboard the early space stations were of a similar nature to camping showers: messy affairs and hard to clean. Like my experience in desert camping, astronauts and cosmonauts were dealing with this challenge in self-care: how to take a real shower in an extreme environment.

In this chapter, you will get to know the state of showering in microgravity, the challenges behind it, and then I propose a new concept of a sophisticated space shower that processes the greywater with a combination of plants inside and outside of the shower itself!

A History of Space Showers

The shower. On Earth, immersing yourself into falling waters is both invigorating for the soul and cleansing for the body. Unfortunately, the few attempts at shower design by NASA and the Russian Space Agency have been inadequate. Insufficient understanding of the surface tension of water made them cumbersome to clean, plus the water and air heating systems were primitive.

As mentioned in previous chapters, the biggest challenge with space showers is surface tension. Without gravity, water tends to stick to itself and other surfaces, making it very difficult to dispose of and clean.

Onboard the International Space Station, astronauts and cosmonauts resort to camping-style techniques to get clean: wet wipes, wash cloths, no-wash shampoo, and lots of cleaning up with dry towels. The experience is not enjoyable, it's just to get clean.

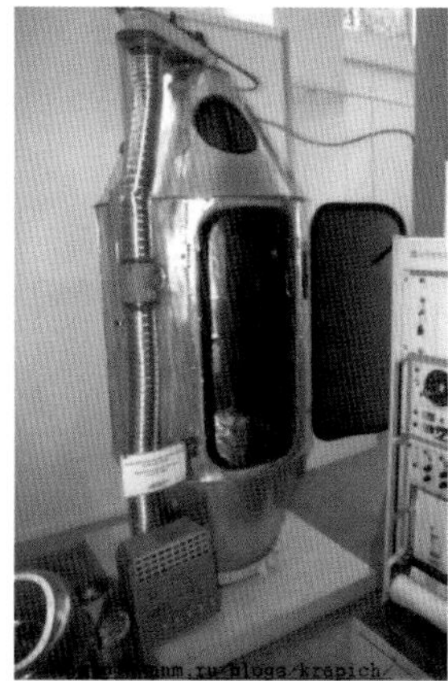

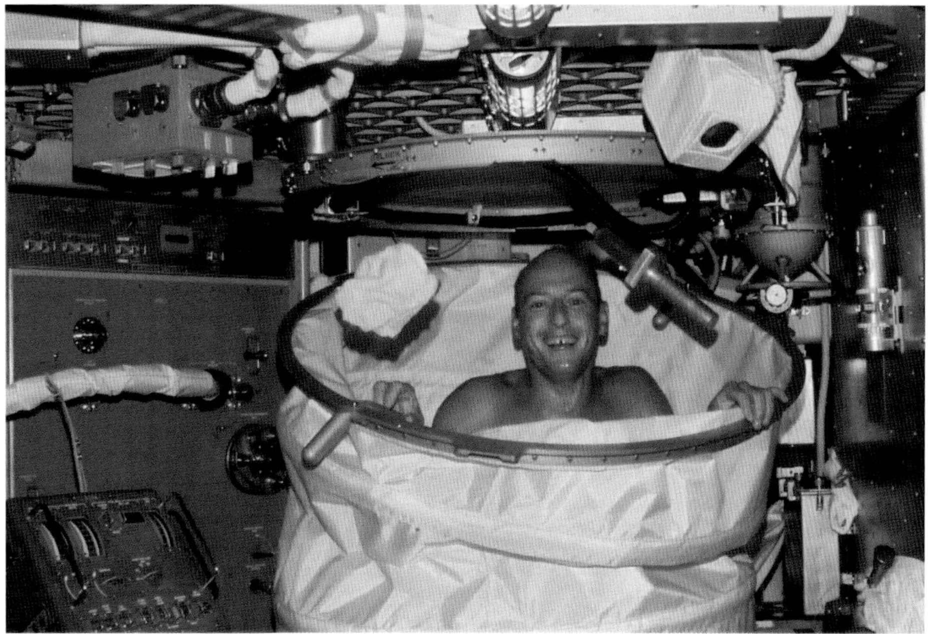

Clockwise: Russian Salyut 7 space shower, MIR Space shower/sauna, NASA's Skylab space shower. Source: NASA.

SKYLAB Space Shower: The most famous space shower is the one that flew on NASA's Skylab space station in the early 1970s.[157] It consisted of a collapsible tube that the astronaut would surround himself in to prevent water leaks. He would use a shower head to wash the body, use soap to clean, and then rinse off. While the showering itself was pleasant, the astronauts reported that the experience of cleaning and breaking down the shower was not. The issues they ran into include:

- The time-consuming nature of cleaning the excess water
- Inconsistency of water temperature

---

157. R. L. Middleton et al., "Design, Development, and Operation of a Zero Gravity Shower," vol. AAS PAPER 74-136 (American Astronomical Society Annual Meeting, Los Angeles, CA, 1974), https://ntrs.nasa.gov/citations/19740059331.

- Inconsistent air temperatures
- Limited amount of water allocated for the shower (3 Liters / six pints)[158]

After one shower experience, the astronauts resorted to wet wash cloths for bathing.[159]

Russian Space Showers: Lesser known are at least two types of showers used on Russian (Soviet) space stations in the 1970's and 1980's. There were attempts on Salyut 3, 6, and 7 to implement a type of shower shaped like a collapsible tube, similar to Skylab. As the Russians were new to dealing with the surface tension issues of water in space, the cosmonauts also had a hard time cleaning up the excess water. The cosmonauts, like the astronauts on Skylab, used towels to dry up the water that clung tenaciously to their bodies and the wall of the shower cylinder.[160]

At least one version of the Russian shower required the use of an air mask. The reason may have been the engineer's concerns about the sticky water covering the cosmonaut's mouth and nose while bathing. From a report:

> *"Fastened at the bottom of the shower are rubber slippers to keep the cosmonaut from floating upward. Above the cosmonaut's head are cellophane bags containing napkins and a towel. The cosmonaut will put a pipe that leads outside the chamber into his mouth and put a clip (like a snorkel) on his nose before turning on the water. Then he opens a package that contains a soap-filled cloth and switches on the water. Water comes out in a fine needle-like spray. The water air aerosol passes through holes in the floor and into the waste container."*

Showering was a complicated process — so much so that the showers, which were expected to be completed by noon, lasted until after 6 p.m.

Like their NASA counterparts, the Russian cosmonauts resorted to wet towels for bathing on most occasions, and only once a month attempted to use the shower system.

ISU/NASA Space Shower Concept

In 1997 a Master of Space Studies student at the International Space University named Susmita Mohanty chose for her project the topic of space showers. She dived deep, deeper than NASA or the Russians did, into the design issues, and with a grant from NASA, developed a concept that was later tested aboard NASA's KC135 zeroG cargo test aircraft. The results of the test showed great promise of improving and simplifying the showering and cleaning process. The results are discussed below.[161]

---

158. William C. Schneider, "Skylab Lessons Learned as Applicable to a Large Space Station, 1967-1974" (Washington, D.C.: National Aeronautics and Space Administration, April 1, 1976), https://ntrs.nasa.gov/citations/19760022256.
159. "Lessons Learned on the Skylab Program" (Houston, TX: LYNDON B. JOHNSON SPACE CENTER, July 18, 1974), https://ntrs.nasa.gov/citations/19760004100.
160. B. J. Bluth and Martha Helppie, "Soviet Space Stations as Analogs, Second Edition" (Washington, D.C.: NASA Headquarters, August 1, 1986), https://ntrs.nasa.gov/citations/19870012563.
161. Susmita Mohanty, "Design Concepts for Zero-G Whole Body Cleansing on ISS Alpha—Part II: Individual Design Project," September 1, 2001, https://ntrs.nasa.gov/citations/20010098604.

Focused on the "Zero-g Whole Body Cleansing" concept, she researched previous concepts, and interviewed several astronauts. Her concept emphasized flow control through vacuuming, temperature adjustment of the stall air and in particular the use of water-repellent (hydrophobic) coatings (she called it the "Lotus Effect") to minimize cleaning work.

Space Shower Concept being tested via parabolic flights onboard a KC135 aircraft. Source: NASA.

ESA Space Shower Concept

In 2008 the shower concept was further developed by researchers Marco C Bernasconi, Meindert Versteeg, and Roland Zenger. As part of the initial studies for ESA's "Space Haven" inflatable habitat, the team defined the components of a space shower, based on previous work from NASA and Russia. They observed that previous space showers "and items related to the support of the crew's wellbeing look like last-minute add-ons to the Space Station layout." They emphasized the important human need for a shower for both hygiene and psychological reasons, and how previous NASA researchers seemed to think "a sponge bath is enough." Their study called out the challenge of cleaning up a temporary shower, plus poor-quality soaps and lack of air and water temperature controls as factors in eschewing showers.[162]

Bernasconi and team pushed back on past criticisms and developed a permanent shower with more advanced vacuum and temperature controls. Plus, they emphasized easier shower cleaning. Unfortunately this project was just a detailed research study and no working prototypes were built.

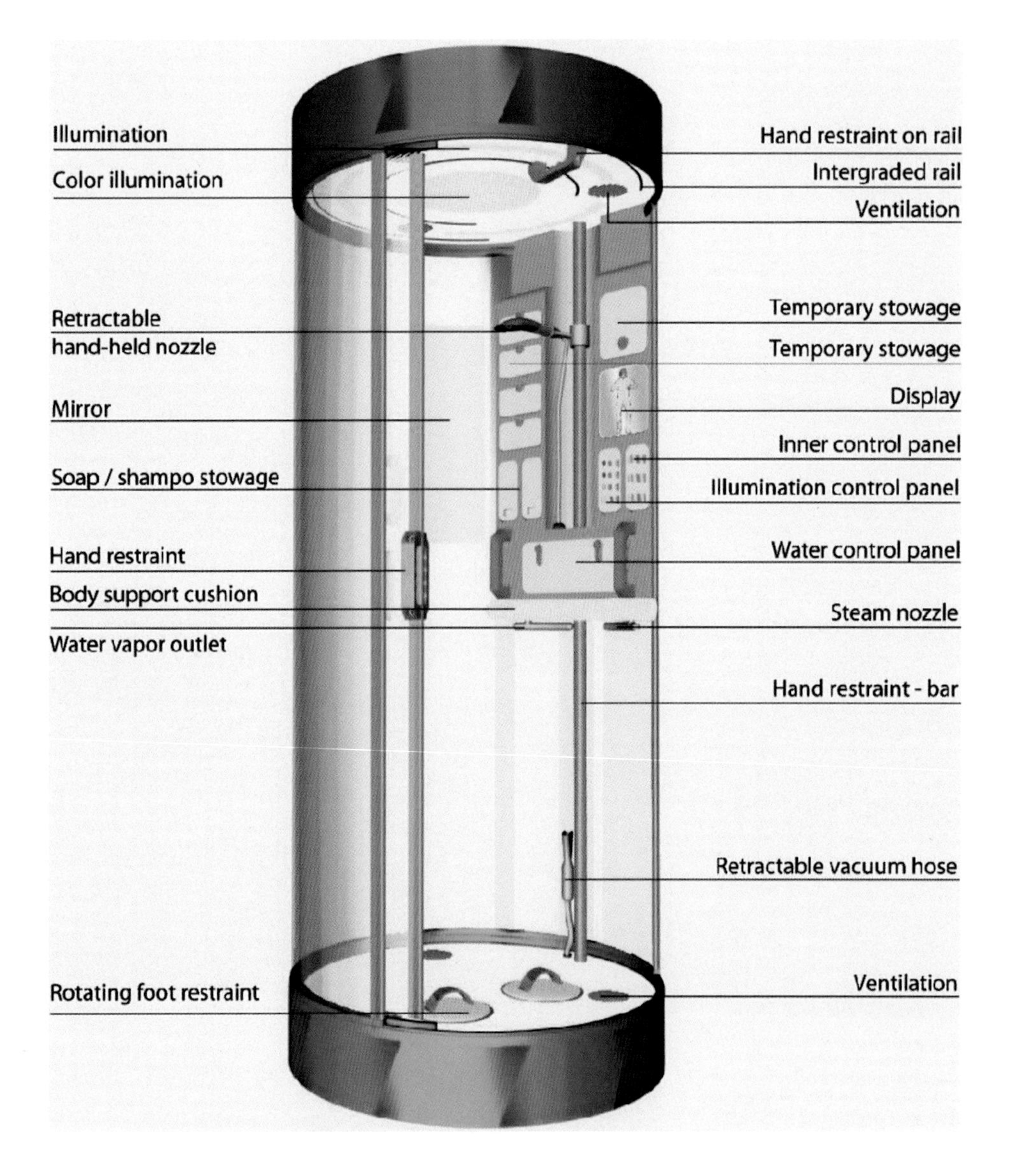

Space Shower Concept developed by Bernasconi et al. for ESA.
Source: Bernasconi, Marco & Versteeg, M. & Zenger, R.. (2008). A Multi-Purpose Astronaut Shower for Long-Duration Microgravity Missions. Journal of the British Interplanetary Society. 172-185.

162. Marco Bernasconi, Meindert Versteeg, and Roland Zenger, "A Multi-Purpose Astronaut Shower for Long-Duration Microgravity Missions" (Valencia (Spain): International Astronautical Congress, October 2, 2006), https://www.researchgate.net/publication/258317776_A_Multi-Purpose_Astronaut_Shower_for_Long-Duration_Mirogravity_Missions.

So beyond a few studies, there has been no serious attempt at building a next generation space shower. If you are going to be living for months or years in a future space station, you darn well need to have a way to take a shower!

What you will read next is my attempt to combine the research above with linkage to the space garden for greywater recycling.

Concept: Shower in the Garden

The new microgravity shower design is based upon the previous research done by Marco Bernasconi's team and the work of Susmita Mohanty with several modifications by me to improve ergonomics and ease of cleaning, and advance a luxury element.

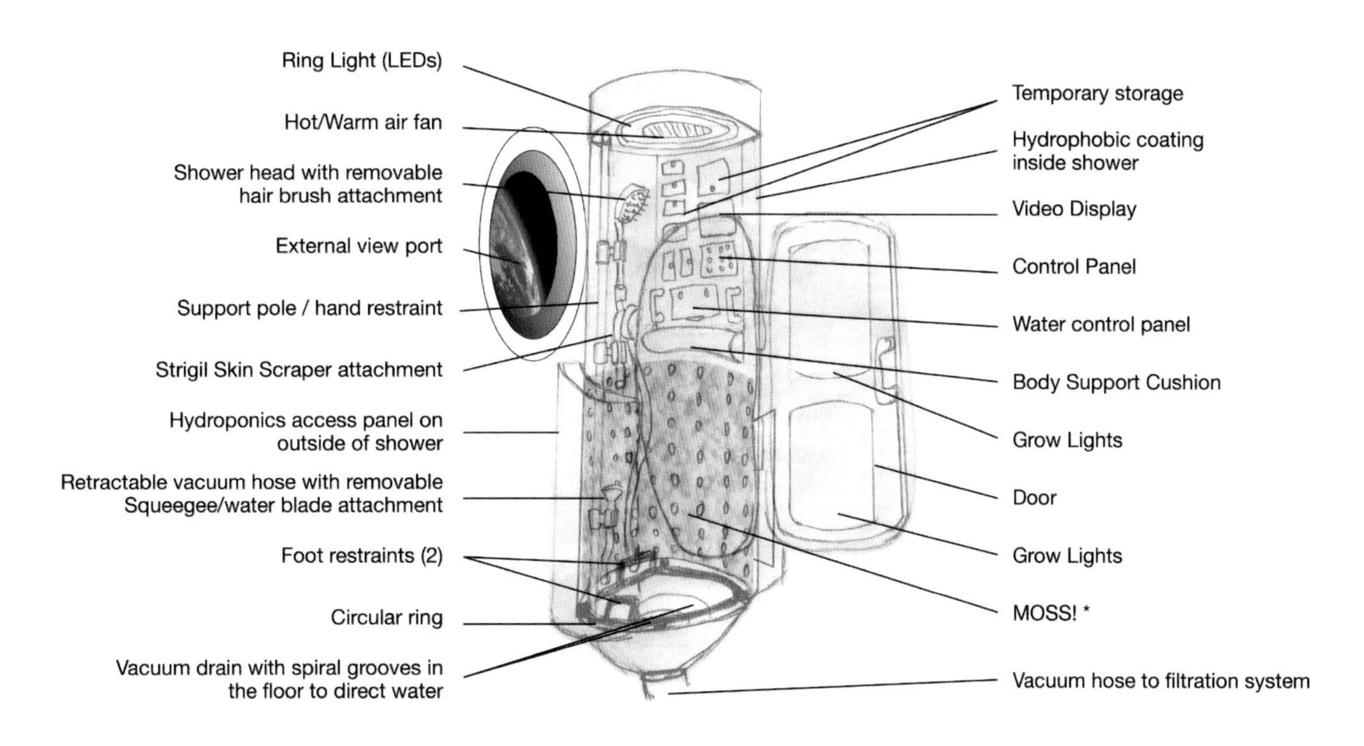

A "Shower in the Garden" concept. Source: the author.

Shower Cylinder

The main part of the shower is shaped like a cylinder, roughly about 2.15 meters (7 feet) tall and 1.22 meters (4 feet) in diameter. This leaves room for drainage and other equipment. There is an LED ring light on the top part of the shower. This light is adjustable in color and brightness. You can control the hot air vent and the blower speed. There are several handles you can grab onto to keep stable.

At the bottom of the shower is a cone with a spiral groove in it that focuses liquids down to a vacuum drain. This vacuum drain is adjustable in speed and power. A circular ring is located near the widest part of the cone; adjustable foot straps are connected to this ring. There are ports for two retractable hoses: one for vacuum and one for water. During the shower, waste shower water is recycled by blasting it with UV-C rays and sent back up to the shower head. Besides saving on water, it is also still warmed from the previous round in the shower.

The cylinder itself is a clear Plexiglass. It is coated with a hydrophobic material so that the water droplets will not stick to its surface.

Shower Controls and Storage

The upper half of the cylinder contains the controls for the shower. This includes water temperature, water flow speed, air temperature, airflow speed, and vacuum speed. There is a digital touchpad for controlling the systems, plus there are manual physical controls just in case the electronics go down.

On the side of the cylinder are several storage containers for shampoo, soap and other personal hygiene items. This control and storage area should only block 1/4 to 1/3 of the plexiglass. The rest of the upper half of the cylinder is clear to allow you to look outside the cylinder.

Moss Garden

There have been numerous studies about the effect of plants in the home to relieve stress and improve well being.[163] In space, where you are far from anything resembling a forest, you need something to remind you of home. A large space garden will serve multiple purposes: provide oxygen, create food, and inspire happiness. I propose that to be both efficient and to improve your sanity, the space shower should be located in the garden! That way the water from the shower can be filtered and sent directly to the garden plants. On top of that, what if you have some green plants to play with...inside the shower!

The space shower design I am proposing includes dozens of small holes on the bottom half of the cylinder for growing moss inside the shower. That way, you can play with the plants and enjoy the shower moment.

Using the technology based on NASA's successful Veggie system, each moss plant will stick out of the hole.[164] Each plant will be connected to a wick which sticks into a pillow-shaped bag of nutrient filled water. This wick will relay water from the bag to the plant. This technique is a common form of hydroponic gardening. On the outside of the cylinder has a maintenance hatch to check on the hydroponics system.

When the shower is not in use, the hatch door on the side of the cylinder is left closed. The air fan and vacuum are continuously running at slow speeds to prevent mold or mildew from forming inside the chamber. The hatch door includes two sets of LED light panels, with which you can control

163. Min-sun Lee et al., "Interaction with Indoor Plants May Reduce Psychological and Physiological Stress by Suppressing Autonomic Nervous System Activity in Young Adults: A Randomized Crossover Study," *Journal of Physiological Anthropology* 34, no. 1 (April 28, 2015): 21, https://doi.org/10.1186/s40101-015-0060-8.

164. Anna Heiney, "Growing Plants in Space," Text, NASA, Kennedy Space Center, July 12, 2021, http://www.nasa.gov/content/growing-plants-in-space.

the brightness and color. These LED lights double as greenhouse lights when they are not being used as shower lights.[165]

Using the Shower

When you enter the shower cylinder, grab onto the long handrail that stretches the entire length of the cylinder. Strap your feet into the footrests at the bottom of the cylinder, and then close the hatch.

How to Take a Shower

For most of us, we are spoiled. All you need to do is turn on the water, and voilá! Instant shower! I'm notorious for taking long showers, as my spouse scolds me for being in the bathroom too long. Hey, what feels better than hot water all over your body as you stretch and get energized for the day?

When you are camping, on the other hand, water is precious, and every drop counts. Rough camping in the wilderness means using a towel to clean your "pits and bits" on occasion. (Unless you can skinny dip into a lake or river nearby that is not freezing).

When traveling with an RV or caravan, and you are "boondocking" in the wilderness without an external water source, you have to rely on your water tanks. My Winnebago camper has a 117 Liter (31 gallon) freshwater capacity, and a (nearly) instant hot water heater. That water is used for cleaning dishes, using the toilet, and keeping clean. (I use a different set of water jugs for drinking). For a family of three, that is enough water for about 3-5 days. Unless you take a shower. The United States Environmental Protection Agency (EPA) states that standard shower heads use 9.5 Liters (2.5 gallons) of water per minute.[166] Multiply that by a five minute shower, that is 47.3 Liters (12.5 gallons)! There goes nearly half of that freshwater tank!

Camping off-the-grid requires a new shower strategy: keep the faucet off as much as possible! Here is an example:

1. Get into the shower chamber.
2. Turn on the hot and cold water.
3. Adjust to your preferred temperature.
4. Get your whole body soaked.
5. TURN OFF THE WATER
6. Shampoo your hair.
7. Turn on the water and rinse shampoo off.

---

165. Fred Sack, "'Moss In Space' Project Shows How Some Plants Grow Without Gravity," *Ohio State News* (blog), January 25, 2005, https://news.osu.edu/moss-in-space-project-shows-how-some-plants-grow-without-gravity/.

166. "Showerheads," Overviews and Factsheets, United States Environmental Protection Agency, October 14, 2016, https://www.epa.gov/watersense/showerheads.

8. TURN OFF THE WATER
9. Put Conditioner in your hair (or whatever you use).
10. Turn on the water and rinse the conditioner off.
11. TURN OFF THE WATER
12. Use soap and a wet washcloth to clean your body.
13. Turn on the water and rinse the soap off.
14. Give yourself a few seconds to enjoy the water.
15. TURN OFF THE WATER

The goal is to get clean, and enjoy the water, with as little water waste as possible.

So how does this apply to space showers? Because you are camping in space just as much as you are camping on Earth. Water conservation is important.

How to Take a Shower in Space

So let's apply the camping shower process to the space shower. Float in, close the hatch and get started! Turn on the hot air to a comfortable temperature, and increase the vacuum speed. Grab the showerhead from its holder.

The showerhead in this concept can be swapped out for a variety of different types of heads. One example showerhead doubles as a hairbrush, with the water streaming out of holes in between the bristles. You can turn on the water, adjust the temperature and pressure, and then press the on/off button on the shower head handle to start getting wet. Because the water will collect together via surface tension, you should start with a small amount of water. Brushing your hair with the water running will get that soapy water deep into your scalp and hair.

Note: You could attempt to take a shower without feet attached to the straps, but the propulsive force of the water coming out the shower head might end up with hilarious results!

You should use glycerin-based soaps and shampoos, which are less harmful to plants. Only a small amount of each is necessary for cleaning the body. I cover the discussion of soaps in the next chapter.[167]

Scraping the Human

When you are done washing and rinsing, now comes the cleaning part. This part of the shower process is quite unusual, due having to deal with microgravity water blobs that stick to your skin.

---

167. Matthew Borden and Adam Dale, "ENY344/IN1248: Managing Plant Pests with Soaps," University of Florida, IFAS Extension, July 21, 2019, https://edis.ifas.ufl.edu/publication/IN1248.

In ancient Greece and Rome, the baths were filled with noblemen and slaves that used an interesting technique for removing oils from their skin. The device was called a strigil.[168] It looks like a tiny sickle with a handle. Because soaps were not available at that time, the Greeks and Romans used olive oil to cover their body. Then after the bath, the oil, dirt, and dead skin would be scraped off with the strigil. (It is also rumored that this icky combination of fluids was sold in the black market for its dubious medicinal benefits, but that is another story for another time.)[169] Modern versions of this device are available today as an exfoliation tool.

The space shower version of the strigil uses a concave blade, not sharp at all, which is attached to a handle. This handle could optionally be attached to the vacuum pump to help suck away the fluids. Your job is to slowly and gently scrape away the soap, water, dead skin and oils that are sticking to your skin. In the past, onboard the Skylab and Salyut/MIR space showers, the astronauts and cosmonauts had to spend a long period of time using towels to wipe off the water from their skin. Using a strigil streamlines the skin cleaning process. One example of a modern version is a crescent moon-shaped blade with a handle called the Esker body plane.[170]

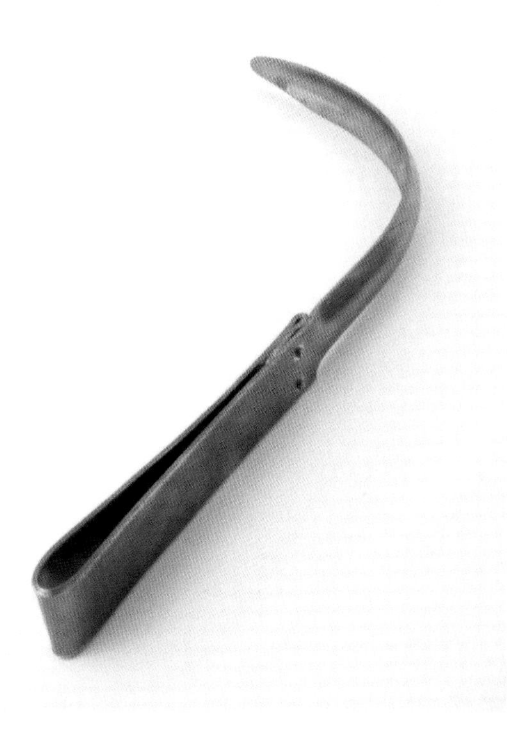

Left: Image on a Greek vase of two bathers holding a strigil. Right: Close up of a replica of an ancient strigil.

168. Marissa Drehobl, Matthew Herman, and Victoria Rduch, "Strigil: Boudoir, Bath and Temple," Amherst College, Mead Art Museum, 2009, https://www.amherst.edu/museums/mead/resources_faculty/faculty/courseproj/boudoir/strigil.

169. Henriette Willberg, "Gloios: Grime, Sweat and Olive Oil," *Ancient Anatomies* (blog), November 21, 2017, https://ancientanatomies.wordpress.com/2017/11/21/gloios-grime-sweat-and-olive-oil/.

170. Emma Trevino, "Your Softest, Cleanest Skin Yet: The Body Plane Tool," *Esker* (blog), June 17, 2020, https://eskerbeauty.com/products/body-plane.

Cleaning the Shower Cylinder

If all goes well the water will have a hard time staying in the shower, for the following reasons:

- The hydrophobic material on the cylinder prevents the water blobs from sticking to it.
- The warm air blowing down from above forces the water down to the bottom of the cylinder.
- The vacuum at the bottom sucks out the water.

For cleaning the residual blobs, you can use another common shower implement: the squeegee! You could use a manual rubber water blade, or one that is attached to a vacuum cleaner.

The adventure of taking a shower with green plants can be luxurious. But like with human skin, surface tension will cause the water to stick to the leaves of the plant. Fortunately, the moss leaves are very tiny, and should be easy to brush off. This act of brushing the plants with your hands and body in the space shower while floating should be a fantastic tactile experience. All the while, the vacuum system will be guiding the water droplets to the drain.

The goal of this new shower system is to reduce the cleaning time and increase the experience. It is unknown how long the whole shower process would take, but your goal is to make it fun and enjoyable while you are in it.

Filtration

Part of the shower design includes reusing the shower water. The vacuum pump can force the water through a filter to remove soap scum, hair, and skin bits, then through a UV-C light chamber to blast apart and potential pathogens, and then be pumped back up through the top of the shower to supplement fresh water coming into the chamber. This reduces the amount of fresh water needed, and should still be warm from the first time through.

After you are done showering, the vacuum and air flow stay on, but the water is then redirected to the plants in the garden.

Plants that thrive on greywater include edible shrubs and vines such as raspberries, thimbleberries, blackberries and their relatives, currants, gooseberries, filberts, rhubarb, elderberry, passion fruit, kiwi, hops, and grapes. Marsh plants like reeds, water hyacinth, iris, and duckweed can also help conserve and treat greywater.[171]

I chose moss in the shower because of its small size and its ability to filter a wide variety of chemicals.[172]

171. Perla Irish, "Marsh Plants That Clean Greywater," *Housesumo.Com* (blog), September 25, 2019, https://www.housesumo.com/marsh-plants-that-clean-greywater/.

172. Kevin Espiritu, "The Benefits of Moss in Your Garden: Add Some Character to Your Garden," *PartSelect.Com* (blog), August 9, 2017, https://www.partselect.com/blog/grow-moss-in-your-garden/.

Linking Showers to Gardens

This easy-to-clean shower design integrated with a hydroponic garden would take the off-world living experience to the next level!

Onboard the International Space Station, astronauts have stated they enjoy working with the plant experiments, and they enjoy the experience of gardening.

Several space farming experiments are currently in progress. Long-term living off-world demands that large scale farming is needed in order for space travelers to survive, if not thrive off-world.

It may be a given that future space habitats have multiple modules devoted to farming, so why not have showers, which use lots of water, located inside the gardens? An exciting garden concept is the Prototype Lunar Greenhouse at the University of Arizona.[173] Their design includes a collapsible cylinder that is expanded in orbit where hydroponic gardens can be developed.

A "Shower in the Garden" becomes a much more enticing prospect when you literally take a shower inside the garden! The shower's greywater can be directed to the plants for filtration and use.

Shower with a View!

The final component of this space shower concept is of course, a window! Though physically having nothing to do with the shower itself, from a design standpoint, it should not be hard to set up the shower in close proximity to a viewport aboard a space habitat.

Imagine this: being naked, soaking in a hot shower while looking out the window at Earth from orbit would be such a peak experience. Would this experience be worth the trip to space?

An amazing view while you are taking a shower! Source: NASA.

---

173. Bob Granath, "Lunar, Martian Greenhouses Designed to Mimic Those on Earth," Text, NASA, April 24, 2017, http://www.nasa.gov/feature/lunar-martian-greenhouses-designed-to-mimic-those-on-earth.

Chapter 5:

Other Creature Comforts

This section is like the kitchen sink, or maybe the junk drawer. All of these concepts will help make long-term living on the homestead comfortable and enjoyable. Some of them, like the space laundry, link into the wastewater processing system mentioned in earlier chapters. The other items are either pure fun, improve comfort, or are badly needed enhancements to the current living situation in space. It is time to move beyond camping and start living a better life off-world.

Volumetric Comfort: When designing your homestead in space, you soon realize you need a lot more room for living, working, and playing, than has ever been done before off-world. Have you ever lived in an apartment that was the size of a closet? I have, unfortunately. It was not a lot of fun. Living in a coffin-sized space capsule can be tolerable for a couple hours or a few days, but sooner or later you need to stretch out. For space habitats and long-term living in them you need some serious room for privacy and freedom of expression. A series of studies shows that a human needs a minimum of 19 m^3 (671 Ft^3) of personal real estate to keep sane for long periods of time.[174] That's like the size of a 3m x 3m x 2.4m (10'x10'x8') storage unit. Keep this in mind while you are reading this chapter and designing your own habitat.

Keeping Clean

At home, I tend to get lazy. With a small child to manage, who has the time or energy to clean the house? It usually gets done when I can't stand the food bits in the rug, the cat fur collecting in the corners, that one stain on the floor you couldn't see until the morning sunlight beaming through the window lands on the floor at just the right angle. At least on Earth, gravity pulls all that stuff to the ground, so you can clean it up in two dimensions.

174. Marc M. Cohen, "Testing the Celentano Curve: An Empirical Survey of Predictions for Human Spacecraft Pressurized Volume," in *SAE International Journal of Aerospace*, vol. 1 (38th International Conference on Environmental Systems, San Francisco, California USA, 2008), 38 pp., https://doi.org/10.4271/2008-01-2027.

In a space station, these same messes are floating, sticking, and moving around everywhere. When I say "everywhere," I mean up, down, left, right, in any direction the little projectiles have momentum to go. The reason certain foods like cookies are forbidden onboard the ISS is their crumbs would fly in three dimensions and be nearly impossible to clean up.

Nearly. Unless of course you design a space station for messy people. Using the homestead analogy, life on a farm can be messy, but there are ways to keep clean. From sweeping the floors, to putting food waste in the compost, to doing the dishes, there are many ways to keep a house a home. It just takes observation and experimentation to figure out how to do it in space.

Concept: Built-in Vacuum Cleaners

Onboard the ISS, there are a handful of air intake vents which astronauts use to suck up finger nail clippings, for example.[175] What if there were a network of air ventilation shafts that could double as vacuum cleaners for debris? The life support system could collect the debris and dump them into a compost system which could "digest" the debris.

As I mentioned before, the lack of convection currents in zero gravity requires artificial currents to be created. That means a second set of air vents are needed to act as the "wind." This more or less resembles the A/C or heating vents in a house, except there are far more of them and they are shaped differently. The wind vents may have temperature and speed controls for each station module, so the crew can manually adjust for personal preference. You can also control the default wind speed and temperature based upon the ambient temperature and air flow in the module. Sensors would be needed in key intersections to measure this.

So now you have the convenience of an ambient air filtration system that does not require you to pull out the Dust Buster. Speaking of which, it may be a good idea to have a portable vacuum cleaner in each room, just in case.

The Robots are Coming!

Rosey the Robot was a popular character in the 1960's TV show "The Jetsons."[176] Not only did she clean the house, she also was programmed with an attitude. Today, we are used to Roombas vacuuming our suburban homes, but they still get stuck in places. Thank goodness they do not have an attitude chip built in like Rosey!

As in science fiction, robots are extremely helpful for doing mundane and dangerous tasks.

NASA is experimenting with rudimentary service robots right now. It turns out astronauts could use some help with their chores, just like many of us on Earth.

---

175. *Chris Hadfield—Nail Clipping in Space*, Canadian Space Agency YouTube Channel (International Space Station, 2013), https://www.youtube.com/watch?v=xICkLB3vAeU.

176. "Rosey the Robot," The Jetsons Fandom Wiki, accessed December 7, 2023, https://thejetsons.fandom.com/wiki/Rosey.

The Astrobee robots are the latest generation of experimental space robots.[177] Each robot measures about one cubic foot and is equipped with six cameras, a touchscreen interface, a speaker, microphone, signal lights, and a laser pointer. The Astrobees flight system operates via two centrifugal impellers that drive air through 12 adjustable nozzles that work in unison. This lets the robot move instantly in any direction and is capable of turning on any axis. The overall maximum thrust for the Astrobees is 0.3 Newtons.

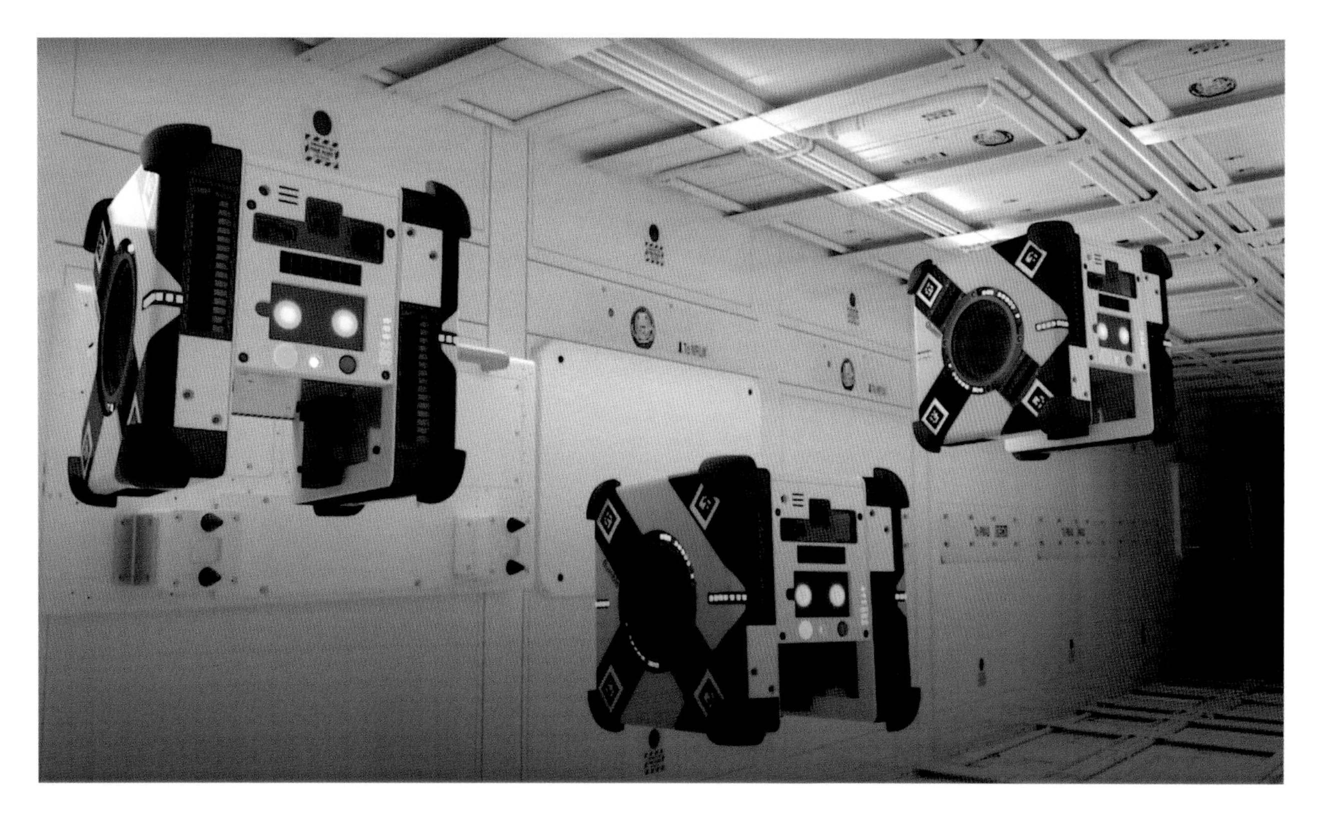

Astrobee Robots. Source: NASA Ames Research Center.

Recently, the Astrobees have been upgraded to become semi-autonomous. The Integrated System for Autonomous and Adaptive Caretaking (ISAAC) software aims to deliver remote and autonomous caretaking during long periods of time when the astronauts are not aboard to perform maintenance, logistics management, and utilization tasks, as well as when communication with ground controllers is limited or simply unavailable.[178] For example, they have been tested to search for lost objects or to detect CO/CO2 leaks.

177. Simeon Kanis, "What is Astrobee?," Text, NASA, November 8, 2016, http://www.nasa.gov/astrobee.

178. Frank Tavares, "Meet ISAAC, Integrating Robots with the Space Stations of the Future," Text, NASA, August 10, 2021, http://www.nasa.gov/feature/ames/meet-isaac.

Concept: Domestic Space Bots

More advanced service robots will make life more convenient on any orbital space yacht or long term off-world settlement. For our homestead design, they are assigned mundane tasks like cleaning rooms, serving drinks and transporting cargo. They also can handle dangerous tasks, such as scanning for radiation leaks, discovering carbon dioxide build up, and detecting and patching vacuum leaks in the station hull.

I actually built a mockup of a mobile service robot in 2005. Little did I know that a few years later a similar robot would be built and become very popular: drones. The power and capabilities of drones in the 2020's are amazing!

The Domestic Space Bot I designed looks similar to underwater drones, or Remotely Operated underwater Vehicles (ROVs). The underwater drones' sleek curved shapes and ducted fan jets are ideal for navigating seaweed and other potential obstacles. Such organic curves in the domestic space bot will come in handy in a space habitat because of the high chance of bumping into walls, people and other things.

It has eight small ducted fans: four pointing in the X axis, two pointing in the Y axis, and two pointing in the Z axis. See the image below:

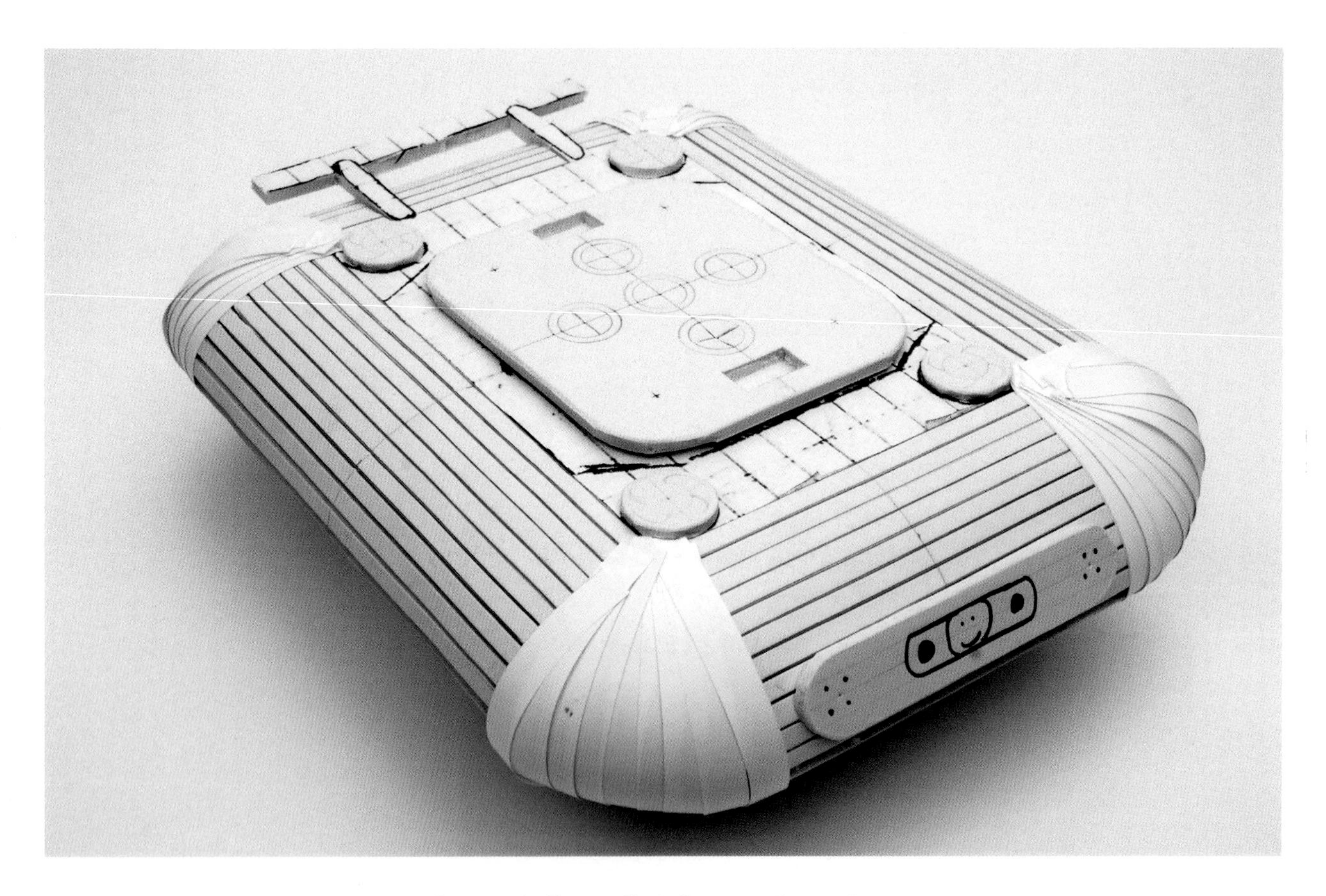

Domestic Space Bot. Source: the author.

In the front, there is a camera, two lights, and multiple sensors: an RFID detector, radiation and chemical detectors, and a CO/CO2 detector. There is a speaker for the robot to talk to people, and microphone for listening or noise detection. There could also be a video display for communication.

There are three mini-gyroscopes in the inside corners of the bot for increased stability in a zero gravity environment. All three should be active when you want the bot to stay in one orientation relative to the task at hand.

In the back of the bot there is a handle. It has two purposes: it makes the robot easier to grab for maintenance, or a place for a wayward passenger to hold onto it while being carried to a safe place such as a ladder, hotel room, or escape pod.

In the center of the bot there is a hole for modules. Because the hole goes all the way through the bot, you can conceive of modules for top or bottom of the bot. Here is a list of example modules:

- Vacuum cleaner
- Manipulator arms
- Serving Tray
- Luggage/cargo carrier
- Science equipment
- Video cameras
- Message displays

The Domestic Space Bots do not have to move fast, but they should be highly maneuverable, and be able to fly into tight spaces. They should have proximity sensors to avoid bumping into things or people. As the ducted fans can be a bit noisy, there should be a sound dampener in the system. Ducted blades with ten or more blades tend to be more efficient.

Permission to Play

Creature Comfort #11: Weight room

The astronauts currently use three specially designed equipment for aerobic exercise and strength training. They are also designed to keep you from floating away. For example, the treadmill requires you to be strapped by the waist to keep from floating away. The Advanced Resistive Exercise Device (ARED) maintains muscle strength and bone density by targeting the major muscle groups, such as the legs.[179] The Stationary bike generates the same aerobic exercise as the Earth-bound version. These machines are a good start, but more exercise options need to be developed.

Human beings are not robots. We cannot always be efficient and focused on work. The challenge with long-term living in space is that sooner or later we have to let our guard down and relax. We want to vent frustrations. We want to push our limits. We want to play. Playing takes many forms and usually requires a lot more real estate that is currently not available in space laboratories like the ISS. If we want to have humans in space for long periods, like months or years, we need to give them room to stretch out or bounce off the walls, literally. I have discovered that there have been some experiments with play on aircraft that do parabolic flights (flying up and down at steep angles), and the results are tantalizing.

Gravity is Just a Habit

The rock band OK GO! Released a stunning music video in 2016. "Upside Down & Inside Out" was an amazing effort of choreography, dance, and play…in an aircraft doing parabolic dives in the air to simulate zero gravity![180] Not only were the band mates playing with balls, candy, pinatas, and paint; the dancers/flight attendants were doing amazing three dimensional choreography, doing spins, twists, and climbing the walls and ceilings. This is a great teaser for the amazing potential of dance and play in space![181]

179. Canadian Space Agency, "Physical Activity in Space," Canadian Space Agency, August 18, 2006, https://www.asc-csa.gc.ca/eng/astronauts/living-in-space/physical-activity-in-space.asp.

180. *OK Go—Upside Down & Inside Out*, 2016, https://www.youtube.com/watch?v=LWGJA9i18Co.

181. *OK Go—Upside Down & Inside Out BTS—How We Did It*, 2016, https://www.youtube.com/watch?v=pnTqZ68fI7Q.

Screenshots of the OK GO "Upside Down & Inside Out" music video.
Source: "OK GO—Upside Down & Inside Out," OK GO's YouTube Channel, 2016.

Kinaesthetics

Jeanne Morel is a dancer, artist and explorer. She is a permanent member of the international dance council at UNESCO and works on the adaptability of the body and the brain in extreme environments and, in particular, in weightlessness.[182] She conducts this art-science research with her partner, the artist and researcher Paul Marlier. Jeanne Morel and Paul Marlier direct ART IN SPACE.[183] They join forces with various scientists, philosophers and astronauts to create an eth-

182. *Jeanne Morel et Paul Marlier—Art In Space*, 2021, https://www.youtube.com/watch?v=E178DbabTd8.

183. Jeanne Morel and Paul Marlier, "ART IN SPACE," ART IN SPACE, accessed October 27, 2022, https://www.artinspace.fr.

ical, poetic and humanistic space quest. Their work is sponsored by the European Space Agency and astronaut Jean-François Clervoy. Jeanne Morel directs the Vulpes Vulpes company and takes dance to unexpected places in order to connect humans, regardless of their physical condition, social condition, age or customs.

Jeanne Morel dancing in a European ZEROG aircraft.
Source: Jeanne Morel et Paul Marlier—Art In Space.

Yoga in Space

Yoga in space is quite interesting since it was designed primarily to counteract the effects of gravity on the body. Granted, without gravity some of the more complex poses (called asanas) can be more easily attained. This could be a great opportunity for yoga experts to develop new types of stretches and positions. Very recently, a popular children's YouTube fitness channel called Cosmic Kids Yoga collaborated with the European Space Agency.[184] Jaime, the host/instructor, teamed up with astronaut Samantha Christoforetti to modify centuries-old yoga poses and apply them to microgravity.[185] Some of the asanas required elastic straps to keep the astronaut in place. Other poses could be done while floating, or loosely tethered via leg straps.

---

184. *Yoga in Space 2—Jaime Visits the European Astronaut Centre I Cosmic Kids Special Project*, 2022, 2, https://www.youtube.com/watch?v=3Npx_ZG95qM.

185. *Astronaut Does Cosmic Kids Yoga IN SPACE!*, 2022, https://www.youtube.com/watch?v=xWPBTqW3ipI.

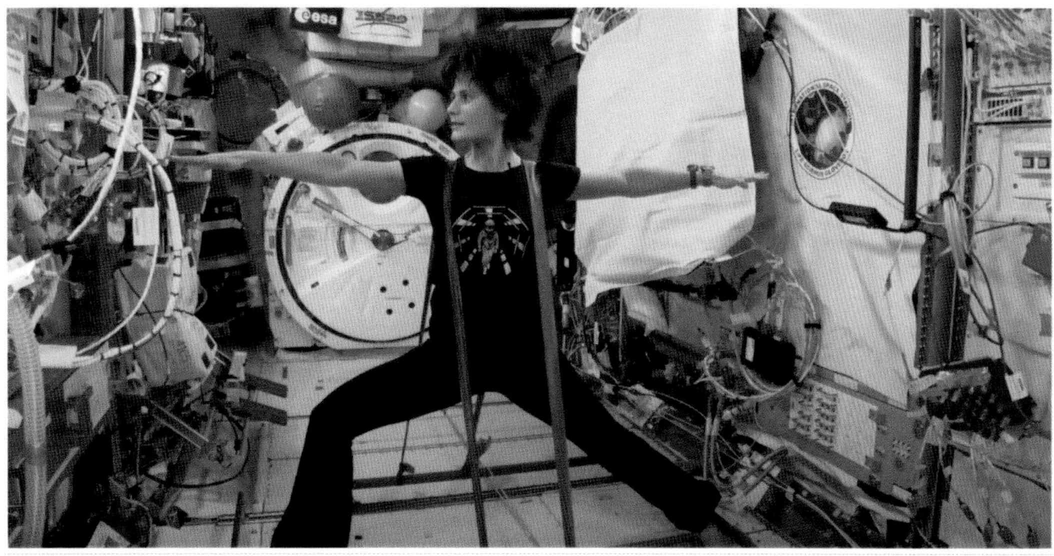

Cosmic Kids Yoga: Top, Jamie and Samantha practice Warrior 2 Yoga pose on Earth. Bottom: Samantha in the same pose in space onboard the ISS. Note the elastic straps needed to keep her "grounded." Source: Cosmic Kids Yoga Youtube Channel.

Space Sports

The Space Games Federation calls itself "the first governing and sanctioning body for competitive sports played in zero or microgravity."[186] They are initiating an "Equal Space" game challenge that combines Science, Technology, Engineering, and Mathematics (STEM) with athletics, art, and media. Their goal is to create "Astroletes": people who have both athletic prowess and astronaut-style tech savviness.[187]

---

186. "Space Games Federation—Sports in Space #EqualSpace™," Space Games Federation, 2022, https://spacegamesfederation.com/.

187. *From Idea To Infinity.* (Space Games Federation, 2019), https://vimeo.com/320018239.

The Space Games Federation wants to develop sports in space. Source: Space Games Federation.

With these few examples, I hope you become inspired to design a new sporting complex off-world! I know I have!

Concept: Space Gymnasium

We need a gymnasium in space. A multipurpose place for sports, dance and for other types of physical activities. In fact, there are several proposals for creating entertainment centers in orbit for the specific purpose of entertainment. Here is my concept:

The space gymnasium would be an inflatable module similar to NASA's TransHab[188] or the Bigelow Aerospace's inflatable module[189] with an entrance on both ends of the cylinder. Motors on each end of the cylinder would include gears to allow movement of the entire cylinder to create artificial gravity. The cylinder can spin at variable speeds to create different levels of gravity. People can experiment with cylinder speeds to develop new types of exercises.

As I mentioned in previous chapters, there are many dangers to the human body by being in microgravity. By having an exercise space with a variable gravity component, you can have a place for your body to readjust to more Earth-like conditions.

By having a very large open air space to exercise in, all sorts of group sports and dance are possible.

---

188. Horacio delaFuente et al., "TransHab: NASA's Large-Scale Inflatable Spacecraft" (2000 AIAA Space Inflatables Forum; Structures, Structural Dynamics, and Materials Conference, Atlanta, GA, 2000), 9, https://ntrs.nasa.gov/citations/20100042636.

189. Jeff Foust, "Bigelow Aerospace Transfers BEAM Space Station Module to NASA," SpaceNews, January 21, 2022, https://spacenews.com/bigelow-aerospace-transfers-beam-space-station-module-to-nasa/.

Laundry that Floats

Creature Comfort #12: Sweatpants

Comfortable clothes are always welcome. Keep in mind, clothing like sweatpants will balloon up because there is no gravity to pull it down. So you might look a bit clown-like wearing such an item. Any clothing can be worn in microgravity, but the fit will be different than the same item worn on Earth. Wanna become a space fashion designer? For more details, check out Barbara Brownie's book "Space Wear" where she dives deep into the relationship of NASA, the fashion world, and how you have to reinvent clothing for space.[190]

After a couple hours of zeroG Dodgeball, it's time to change clothes, and put the dirty items into the wash. The only problem is, there is no washer and dryer for space use. How do the astronauts deal with stinky clothes? Here are the current options:[191]

- **Option One:** Wear It Again (as many times as you can until you cannot stand the smell)
- **Option Two:** Turn It Into A Shooting Star (aka dump it into space to burn up in Earth's atmosphere)
- **Option Three:** Grow Plants With It (substitute for dirt)
- **Option Four:** Feed It To Bacteria (theoretically, some bacteria will help digest the fabric)

I am not joking. I also checked with the Canadian Space Agency's website on astronaut hygiene. All it said was:

> *It is impossible to wash clothes on board the ISS! Quite simply, it would take too much water. The astronauts therefore wear their clothes until they are too dirty and then throw them out. All ISS waste burns up in the atmosphere on re-entry.*[192]

This is not acceptable, especially for people living for months or years in our homestead in space. This got me to thinking about the fundamentals: What are dirty clothes? How can you clean them?

---

190. Barbara Brownie, *Spacewear: Weightlessness and the Final Frontier of Fashion*, 1st ed. (London: Bloomsbury Visual Arts, 2019), https://doi.org/10.5040/9781350000353.

191. John Ira Petty, "Astronauts' Dirty Laundry," Feature Articles, NASA News (Brian Dunbar, April 10, 2003), https://www.nasa.gov/vision/space/livinginspace/Astronaut_Laundry.html.

192. "Personal Hygiene in Space," Canadian Space Agency, August 18, 2006, https://www.asc-csa.gc.ca/eng/astronauts/living-in-space/personal-hygiene-in-space.asp.

Dirt and Stains

What makes clothing dirty? That seems like a silly question, but when you actually look at examples of dirt and stains and odors, you realize there is some chemistry and biology in the process of getting dirty. On Earth, clothes can get dirty from just about anything: body oil, grease, ketchup, paint, and nearly anything liquid in your kitchen. Some liquids, like water or alcohol, tend to dry up quickly and disappear. Others stick like glue (it could be glue) and require some alternative chemical to remove the stain. The tough stains tend to be body oils. There are legions of books and magazine articles about home remedies for removing common, and not so common stains.[193]

The Great Stink

Besides stains, body odor on clothing is the worst. Even with a clogged up nose in space, you may still notice a whiff of the funky smell from yourself and others. Body odor happens when bacteria on your skin come in contact with sweat. Our skin is naturally covered with bacteria. When we sweat, the water, salt and fat mix with this bacteria and can cause odor. The odor can be bad, good or have no smell at all. Factors like the foods you eat, hormones or medications can affect body odor. Every time you sweat, there's a chance you'll produce an unpleasant body odor. Some people are more susceptible to foul body odor than other people.[194] So as the saying goes, "you are what you eat!" NASA tries to counteract this by selecting activewear made from materials with antibacterial properties, such as merino wool. It is 'anti-stink' to a point, but after 30 days of continuous wear, any clothing is going to smell icky.

The Power of Soap

The variety and formulations for soap are astounding.[195] It should be listed in the top ten greatest inventions, next to the steam engine and the Internet. There is soap to clean skin, clean dishes, clean clothes, clean cars, and many more. Some soaps are all-purpose, meaning they can clean many of the aforementioned things, if it is diluted properly. In a resource-limited environment like our space station, this kind of soap would be the best option. The fascinating thing about soap is that part of its structure attracts oils, while the other part of it attracts water. Very little soap is needed to break down oils and to kill bacteria. One of my favorites is Dr. Bronner's Castile Soap, which I discovered from some friends at Burning Man and have been using it for camping and at home for years! Their website even has a dilution cheat sheet for dozens of uses![196]

---

193. Virginia Friedman, Melissa Wagner, and Nancy Armstrong, *Stains: A Spotter's Guide* (New York: Barnes & Noble Books, 2005).

194. "Body Odor: Causes, Changes, Underlying Diseases & Treatment," Cleveland Clinic, March 4, 2022, https://my.clevelandclinic.org/health/symptoms/17865-body-odor.

195. Ferris Jabr, "Why Soap Works," *The New York Times*, March 13, 2020, sec. Health, https://www.nytimes.com/2020/03/13/health/soap-coronavirus-handwashing-germs.html.

196. Lisa Bronner, "Dilutions Cheat Sheet for Dr. Bronner's Pure-Castile Soap," Dr. Bronner's, February 5, 2023, https://info.drbronner.com/all-one-blog/2023/02/dilutions-cheat-sheet-dr-bronners-pure-castile-soap/.

Sending out to the Cleaners

Some say that dry cleaning is more effective than washing with soap. Certain materials such as silk cannot be cleaned properly using standard soap and water techniques. The dry cleaning chemical choice is perchloroethylene (tetrachloroethylene), which is effective in removing stains.[197] The chemical is put into a washing machine-like device and is used instead of water and laundry soap. Like with water, you still have to deal with blob management of these chemicals, plus how would you dispose of them?

Then there are the other aspects of the dry cleaning process, such as steaming and pressing the clothes to make them wrinkle free. This brings up an interesting question: how nice do your clothes have to look in space? Since you will be basically living in a glorified farm, who is going to care that you do not look like a fashion model? While we are at it, there's going to be a limited amount of real estate on the space station for such equipment. Also, the chemical is extremely specialized and not as multi-functional as soap. Also, it is very toxic. So in my opinion, dry cleaning in space is not worth the effort.

Going for a Spin

In 2021, at the NASA Glenn Research Center, six groups of college interns were given a challenge to design a space washing machine.[198] They were only given a week to do the project, so the results were more conceptual than applicable. The winning team, The Agitation Rollers Team, took the lead with the "SLUSH (Screw-Like Undirtying Spinning Hardware)." (I did not invent these names). They proposed a spring loaded corkscrew device that spins and pitches dirty clothes in microgravity. Other teams proposed magnets and capillary action to squeeze out water, vacuum hoses and rollers with squeegees, or a chamber coated with titanium dioxide, ultraviolet light and ozone gas to kill bacteria. While all of these concepts are quite creative, none of them went beyond the concept stage to be tested. They all seem to address different aspects of the same problem: how to clean clothes. I really would like to see a combination of some of these ideas put together in a demonstration model. What would you do if you had to design a washing machine for space? Where would you start?

Speaking of which…

Tide in Space

Proctor and Gamble designed for NASA a fully degradable detergent called "Tide Infinity." It is custom designed for space to solve malodor, cleanliness, and stain removal problems, while also being suitable for use in a closed-loop water system. From their press release:

---

197. "Wet Cleaning vs Dry Cleaning | Electrolux Professional," *Electrolux Professional Global* (blog), June 21, 2023, https://www.electroluxprofessional.com/wet-cleaning-vs-dry-cleaning-pros-and-cons/.

198. Kelly Sands, "NASA Glenn Interns Take Space Washing Machine Designs for a Spin," Text, NASA Glenn Research Center, July 28, 2021, http://www.nasa.gov/feature/glenn/2021/nasa-glenn-interns-take-space-washing-machine-designs-for-a-spin.

> *"This innovative laundry solution will advance cleaning solutions for resource-constrained environments like deep space and water scarce areas on Earth. Tide Scientists are working with NASA to push the bounds of resource efficiency, uncovering learnings with practical applications for the future of laundry on Earth."* [199]

Like I've been saying in this book, if you can design for space, you can apply it to Earth. I respect the intentions of P&G for developing this detergent. Hopefully it will be made available for us humans on Earth soon. The key words here are "suitable for use in a closed-loop water system." Which means the detergent enzymes (used to kill the bacteria and eliminate smells) will themselves break down into harmless molecules which will be easy to separate from water onboard the ISS's life support system.[200] In December 2021, the Tide Pens with the Tide Infinity detergent were flown up to the ISS via SpaceX CRS-24 mission.

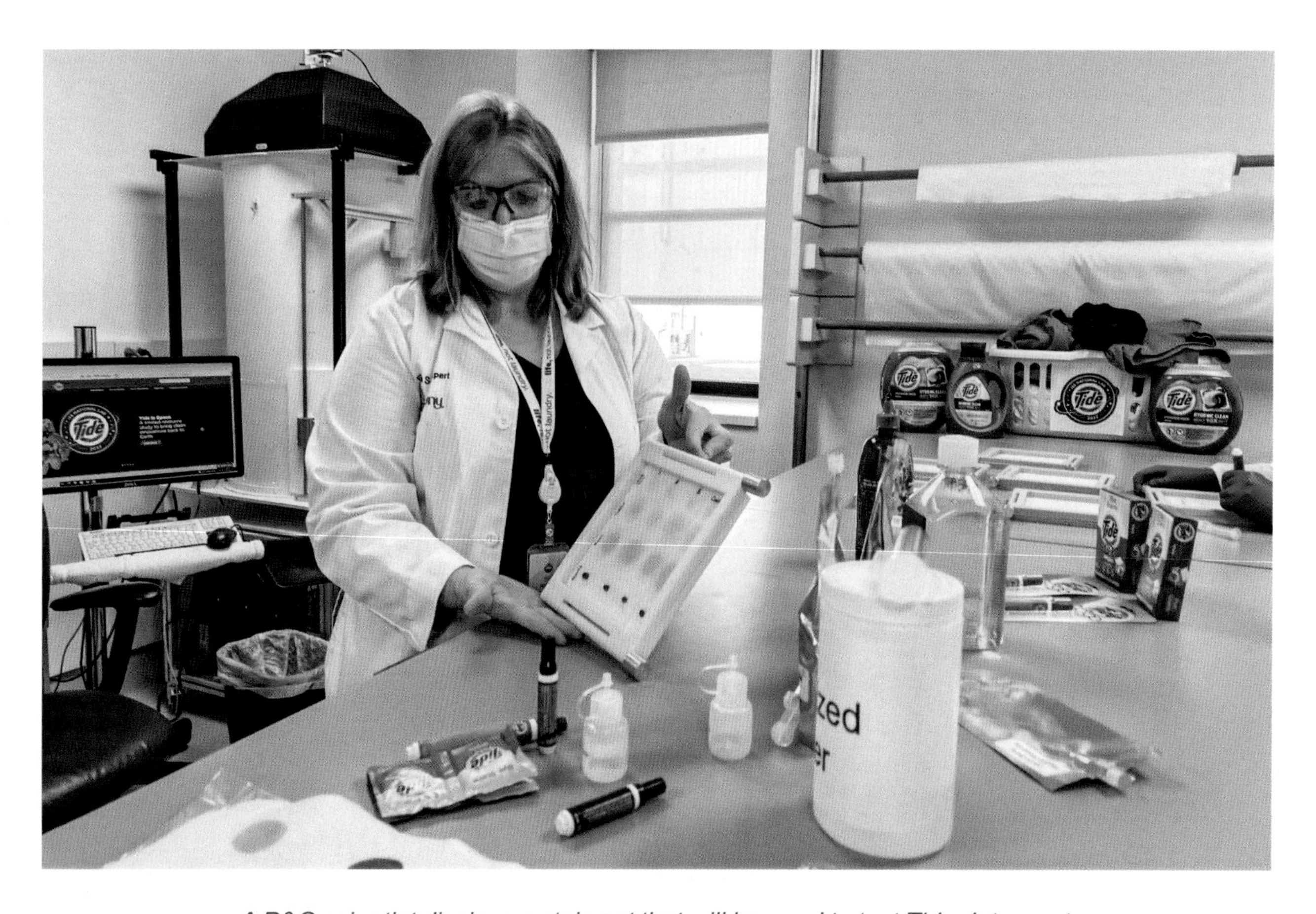

A P&G scientist displays a stain set that will be used to test Tide detergent solutions on board the ISS. Source: NASA and P&G.

199. "P&G's Tide Is Headed to Space with NASA and SpaceX!," P&G, December 20, 2021, https://us.pg.com/blogs/tide-is-headed-to-space/.

200. "P&G Telescience Investigation of Detergent Experiments," NASA Space Station Research Explorer, December 10, 2021, https://www.nasa.gov/mission_pages/station/research/experiments/explorer/Investigation.html?#id=8595.

Laundry Sheets

A new concept which I recently discovered is called laundry sheets. Instead of having big wasteful bottles of laundry detergent, you can use these rectangular sheets of detergent to put into the washing machine.[201] The sheets dissolve instantly and do their thing with cleaning the clothes. It seems to be a hot new product in the eco laundry market. This is an ideal solution for space travel, because they're thin, light, and easily shippable to space. Now we just need a machine to put them in.

Washing Machine in a Bag

Going back to camping yet again as a useful analogy, there are things called *clothing wash bags*. The device goes like this: you put your dirty clothes into the plastic bag, add some water, and add a small amount of laundry soap (a little goes a long way). Seal the bag, and deflate it with a built-in valve. Then you rub the clothes together inside the bag. The clothing rubs against a flexible inner washboard to help with the scrubbing process. After scrubbing for about 60 seconds, open the bag, pour in some warm water to rinse the contents, and then dump out that extra water. Squeeze out the remaining water that's in the clothing. After that, you can hang the clothing to dry on the line or use towels to roll up the clothes to absorb all the water.[202]

A modification of this very simple device could be used right now on the ISS. Intake and outtake valves could be installed in the device to work directly with the space station's water system and go straight to the life support system for processing.

This wash bag system makes perfect sense for individuals, but is there a way you can scale it up for a group of people?

Concept: Space Laundromat

We are spoiled with modern conveniences such as the washing machine. We no longer have to wash our clothes in the river and beat them with rocks!

Joking aside, in a space habitat, water is precious. Every drop is important and currently you cannot waste it on things like washing clothes. For the astronauts on the International Space Station, they get regular shipments of fresh clothing, because they wear each of their clothes for up to 30 days. They then dump them into a cargo vehicle to burn up in the atmosphere. This is not sustainable. There must be a way that people can wash and clean their clothes.

Back to my concept of deep integration with a greenhouse in space. If certain chemicals can be found that are safe for cleaning clothes and processed through plants, there may be a possibility of creating a washer/dryer in a space habitat that also can have wastewater recycled. Have a dedicated section of the garden using plants that are used to filtering out toxins, and you could recycle the same water for washing clothes.

---

201. Kevin Hinton & Ryan Mckenzie, "Tru Earth," Tru Earth, accessed November 2, 2022, https://www.tru.earth.

202. "Scrubba Wash Bag—Travel Washing Machine," The Scrubba Wash Bag, accessed November 2, 2022, https://thescrubba.com/products/scrubba-wash-bag.

Let's break down the challenges:

- Dirt, grease, and bacteria all have to be dealt with
- Skin oil tends to discolor clothing and produce smells
- Water is limited
- Water acts like a blob so it will be persistent and stick to things
- Detergents have to not only clean the clothes but easily break down and be separated from the water.
- Wide variety of stains that could happen on the clothes. Some are very persistent.

Using current washing machine concepts, the water most likely would simply glom onto the clothing and the walls of the washer cylinder. It would look like one giant blob of water with clothing inside. because of this, not very much water would be needed, since the water will try to cover the entire surface area of the clothing. By adding blades or slosh baffles inside of the cylinder, you could control the state of the water and the clothing. When the clothing goes through a rinse cycle, the cylinder could spin at high speeds and the blades would hold the clothing in place. (The potential gyroscopic forces generated from the spinning cylinder could be an issue affecting space station stabilization. More research is needed here.) Use of a squeezing plate and a suction or vacuum inside the cylinder could remove most of the water from the clothing. When in dryer mode, the spin could increase and the heat could increase to evaporate the water from the clothing. NOTE: you may want to add UV-C lights to help kill germs, but only in moderation. This light is known for breaking down the materials of the fabric. All the water, dirt, debris and cleaning chemicals would be sucked into the hydroponic garden. The garden plants would absorb the chemicals and dirt and process them.

Advanced Concept: Grow-Your-Own Clothing

This topic enters science fiction territory. Bear with me. If people live in our homestead in space long enough, years for instance, sooner or later their clothes will wear out. You may be able to order new clothes from Earth a few months in advance and a cargo shuttle will arrive and deliver them (Amazon delivers to space? Guess who owns the Blue Origin space company, folks! The idea is not so far-fetched). What if there was another option, so you don't have to wait months for that cargo ship?

What if you could grow or print or robotically sew your own clothing? What if that fabric is repairable and reusable using advanced 3D printing? 3-D printed clothing is becoming more of a possibility with the emergence of fashion houses, such as ZER in Spain and the 3-D printed BIOSUIT designs by Dava Newman's MIT space suit team.[203] [204] Pretty soon it will be possible for someone to simply

203. "ZER Era (3D Printed Fashion House)," Zer Era, accessed December 11, 2022, https://zereraofficial.com/collections/all.

204. Sarah Beckmann, "Dava Newman Presents 3D Knit BioSuit™ at 2022 MARS Conference," MIT Media Lab, March 30, 2022, https://www.media.mit.edu/posts/dava-newman-presents-3d-knit-biosuit-at-mars-conference/.

melt their old clothing down to the original materials, typically plastic, cotton, or wood, and then put it into a specially designed 3-D printer that can create threads, fabrics, and full outfits.[205]

There is a fashion industry movement called "bio-fabrication" where you can grow fabric using mycelium, a type of fungus. (EEWWW! That's what I said as well when I first heard about this). The resulting fabric is leather-like in feel, non-toxic, and is water and fire-resistant. Current techniques require 2-4 weeks to grow enough fabric.[206] A new material with the brand name of Reishi can be treated and manufactured like leather, that supposedly outperforms leather in strength and matches leather in durability and appearance.[207]

What does this mean for our space homestead? It means you could one day potentially grow your own clothing! When it wears out, you can patch it, or throw the whole thing into the space station's life support system to compost it down as food for the space garden. There could be a mycelium factory that grows more fungus via fermentation, possibly around a body form that is adjustable to the current shape of your body, and in a short time voila: new clothing. Bio-fabrication is still in the experimental phase right now, but it has interesting potential for long term clothing options in space habitats in orbit and on other worlds.

Sleeping

After a hard day's work, it's time to get some shut eye. Unfortunately, sleeping in space is quite a bit different than sleeping on Earth. First of all, without gravity you float in a fetal position: your arms and legs start floating away from your body. If you close your eyes and let go of everything, you slowly float around your cabin. Sooner or later you'll probably bump into something. Onboard most spaceships this is not practical or safe. There's lots of sensitive equipment, wires and things you could bump into, and you could possibly get hurt. For astronauts, it is very hard to get a good night's sleep in space.[208]

Also, it's very noisy up there. At least onboard the ISS, the noisy machinery of the station is a constant hum in the background. You could probably design sleeping quarters with better insulation to reduce the noise, but it would probably not completely get rid of the ambient hum of the machinery that keeps the space station going. Another issue is the light. Unless you design a cabin that can completely block all light you will probably get some of it in your face and distract you. The craziest thing you will encounter is cosmic rays. Several astronauts have reported that when they're sleeping

205. Samir Ferdowsi, "How Far Away Are We From Downloading Our Clothes?," Refinery 29, April 29, 2021, https://www.refinery29.com/en-us/3d-printing-fashion.

206. Charlie Bradley Ross, "Fabric Made From Fungi," *The Sustainable Fashion Collective* (blog), August 24, 2016, http://www.the-sustainable-fashion-collective.com/2016/08/24/fabric-made-fungi.

207. Jessica Wolfrom, "When Fashion Is Fungal," Washington Post, August 31, 2020, https://www.washingtonpost.com/climate-solutions/2020/08/31/fashion-musrhooms-mycelium-climate/.

208. Laura K. Barger et al., "Prevalence of Sleep Deficiency and Hypnotic Use Among Astronauts Before, During and After Spaceflight: An Observational Study," *The Lancet. Neurology* 13, no. 9 (September 2014): 904–12, https://doi.org/10.1016/S1474-4422(14)70122-X.

they notice sparks or flashes in the retinas.[209] As far as scientists can tell, this is because cosmic rays are blasting through their bodies and sometimes their eyes.[210] The shockwave affects the optic nerve and causes the flashing effects. There is no good way to block this radiation except through thick walls or a Star Trek-style force field.[211]

I almost forgot: the lack of airflow in the tiny cabin. If the bedroom cabin is sealed too well, there will be no airflow and slowly your oxygen will be used up and replaced with carbon dioxide. Not good.

In the past, cramped spacecraft forced astronauts to sleep in their cockpit. On board the U.S. and Russian space stations, the astronauts and cosmonauts have private sleeping quarters. What about future space stations? What if you have visitors who are a married couple? Can you design a honeymoon suite in space? What would it look like? How would it function? What if you had a family onboard: Parents and children? What kind of accommodations can you design for that situation in a microgravity environment? As a father of a young child, I imagine my son would enjoy literally bouncing off the walls all the time.

This is the ultimate creature comfort challenge: with the design constraints listed above, can you design a bedroom where a person can get a good night's sleep?

Light is a Stimulant

Creature Comfort #13: Indirect lighting

Much of the ISS interior is filled with indirect lighting so that you are not blinding or distracting people with harsh direct lights.

In a report published in 2016, researchers studied the sleep patterns of astronauts on both the Space Shuttle and the ISS. [212] They learned that the body's internal circadian rhythm gets disrupted in the 90 minute day/night cycle of orbiting around the Earth. Circadian rhythms are physical, mental, and behavioral changes that follow the 24-hour cycle that we have known since birth. These natural processes respond primarily to light and dark and affect most living things, including animals, plants, and

209. Nathaniel Scharping, "What Keeps an Astronaut Awake at Night? Cosmic Rays," Discover Magazine, December 19, 2017, https://www.discovermagazine.com/the-sciences/what-keeps-an-astronaut-awake-at-night-cosmic-rays.

210. G. G. FAZIO, J. V. JELLEY, and W. N. CHARMAN, "Generation of Cherenkov Light Flashes by Cosmic Radiation within the Eyes of the Apollo Astronauts," *Nature* 228, no. 5268 (October 1, 1970): 260–64, https://doi.org/10.1038/228260a0.

211. Joseph John Bevelacqua and Seyed Mohammad Javad Mortazavi, "Commentary Regarding 'on-Orbit Sleep Problems of Astronauts and Countermeasures,'" *Military Medical Research* 5, no. 1 (October 30, 2018): 38, https://doi.org/10.1186/s40779-018-0185-2.

212. Steven W. Lockley and George C. Brainard, "Lighting Effects" (NASA Technical Reports Server, June 7, 2016), https://ntrs.nasa.gov/citations/20160006727.

microbes.[213] Studies of astronauts who flew between 2001-2011 on the space shuttle and 2006-2011 onboard the ISS showed that astronauts slept considerably less in space than they did on Earth.[214] This light study, known as the Lighting Effects study, coincides with a lighting 'makeover' on the ISS in 2016. The fluorescent lights on the station were replaced with a new system of solid-state light-emitting diodes (LEDs). Not only are the LEDs more energy-efficient and safer, they can literally shed all sorts of light on the study topic. From the study:

> *"Light has a number of effects on our sleep and circadian rhythms; is a natural stimulant and can improve alertness and performance, and also help reset the 24-hour clock when it gets out of sync. The human eye contains a light-sensitive protein called melanopsin, different from the rods and cones that we use to see, which detects light in the eye and mediates this effect. Melanopsin is most sensitive to short-wavelength blue light and so by increasing or decreasing the proportion of these blue wavelengths in white light, we can enhance alertness, or promote sleep, respectively. NASA has developed a multi-LED lighting system to take advantage of these light effects. The system can provide millions of different light spectra. We're not making the ISS into a disco, but we are going to use three different light settings. We'll use a general light setting that provides a good light to see by during normal work, a higher-intensity blue light enriched setting that elevates alertness and can better shift the circadian clock when needed, and a lower-intensity blue wavelength-depleted 'pre-sleep' setting to calm the brain and promote sleep. We will be studying the impact of these lights in future missions."*[215]

(I kinda want to see a disco ball with multi-colored lights and dancing astronauts, but that's just me). Anyways, the idea is that warmer colors such as orange indicate slowing down and preparing to rest, whereas colder blue colors focus the person and keep them awake. You probably noticed this feature in your smartphone, where the screen will automatically reduce the blue colors from the screen in the evening to make it easier on your eyes, and in theory remind you to turn off social media and get to bed.

Bulging Eyeballs and Vacuum Pants

Another challenge with microgravity is the fact that extra blood flows to the upper body and puts pressure on the brain and the eyes. NASA calls this disorder Spaceflight-Associated Neuro-ocular Syndrome, or SANS. Researchers at the UT Southwestern Medical Center are developing what could be called vacuum pants: they suck excess fluids from the upper torso to your legs.[216] The Russians

---

213. "Circadian Rhythms," National Institute of Health (NIH), National Institute of General Medical Sciences (NIGMS), May 4, 2022, https://nigms.nih.gov/education/fact-sheets/Pages/circadian-rhythms.aspx.

214. "The Power of Light | Science Mission Directorate," NASA Science, The Power of Light, December 13, 2016, https://science.nasa.gov/news-articles/the-power-of-light.

215. *ScienceCast 234: The Power of Light* (NASA Image and Video Library: NASA HQ, 2016), 234, https://images.nasa.gov/details-234_PowerOfLight.

216. "High-Tech Sleeping Bag Could Solve Vision Issues in Space," UT Southwestern Medical Center Newsroom, December 13, 2021, https://www.utsouthwestern.edu/newsroom/articles/year-2021/high-tech-sleeping-bag.html.

have a similar device known as a Chibis suit.[217] They work like a household vacuum cleaner to suck the body's fluid into the pants, towards the bottom of their feet, and expand veins and tissues of the lower body. By sucking blood and other body fluids back to the lower body, you can reduce the swelling and pressure in the face and head. Imagine spending months in space, slowly going blind, and having to wear rubber vacuum trousers to bed? This is another incentive to build variable gravity space stations, in my opinion.

Another option is to wear a skinsuit. During the 2023 AXIOM Space AX-2 commercial flight, the private astronauts tested a new type of suit to counteract the effects of microgravity.[218]Called the The Gravity Loading Countermeasure Skinsuit, it is sponsored by the Massachusetts Institute of Technology's (MIT) Media Lab's Space Exploration Initiative. It is a body suit that covers the entire body except for the arms. The goal is to simulate the force of gravity on the body to reduce muscle and bone atrophy. A body suit that is less bulky than the vacuum pants to me seems like a more comfortable option, plus you can actually float around and do work.[219]

Noisy as Heck

Creature Comfort #14: Silence

While outer space is itself very quiet, the International Space Station is a rather noisy place. It is nearly impossible, even in the astronauts' private quarters, to get much quiet. They tend to wear earplugs a lot. Future space station designs will have to address the issue of vibrations and noise isolation systems to improve passenger and crew comfort.[220]

217. Le Warren, "Rubber Vacuum Pants That Suck," NASA Blog: A Lab Aloft (International Space Station Research), June 2, 2015, https://blogs.nasa.gov/ISS_Science_Blog/2015/06/02/rubber-vacuum-pants-that-suck/.

218. Leonard David, "Having 'Skinsuit' in the Game: Managing Microgravity," *Leonard David's INSIDE OUTER SPACE* (blog), May 20, 2023, https://www.LeonardDavid.com/having-skinsuit-in-the-game-managing-microgravity/.

219. James M. Waldie and Dava J. Newman, "A Gravity Loading Countermeasure Skinsuit," *Acta Astronautica* 68, no. 7 (April 1, 2011): 722–30, https://doi.org/10.1016/j.actaastro.2010.07.022.

220. James L. Broyan, Scott M. Cady, and David A. Welsh, "International Space Station Crew Quarters Ventilation and Acoustic Design Implementation" (International Conference on Environmental Systems, Barcelona, 2010), https://ntrs.nasa.gov/citations/20100017014.

The ISS open cabin can reach a constant noise level of 80 dB.[221] That's like keeping the vacuum cleaner on, forever! About one quarter of the astronauts would use earplugs, and about three-quarters of them take sleeping pills to try to get any sleep. It's similar to a noisy airplane in flight. What can be done to manage the noise?

Pain in the Neck

During Shuttle missions, approximately one-third of crew members reported that the most important part of the body to be concerned with during sleep is the head. Why? Because it can nod, tilt, and twist during sleep. Remember, your body is floating, even if it is constrained to a sleeping bag. That means your head is also floating around while you are sleeping in microgravity. Weird huh? Astronauts complained of soreness in the neck the next day. Traditional pillows don't work, since your head is floating around. A neck brace sounds like torture. So what is an alternative?

Seeking Fresh Air

Creature Comfort #15: Heating & Air Conditioning

Temperature Control in a room can be an issue. People tend to have personal preferences of their ideal temperature, and the preferences between men and women can be dramatically different. Onboard the ISS, the ambient temperature is around 27°C (80°F) which can be toasty for some people.[222]

Even NASA engineers struggle with air flow issues. When they designed the sleeping pods called "crew quarters" for the ISS, they struggled to incorporate air intake and exhaust fans that would provide enough circulation, and yet be not so loud from the fan motors.

Concept: Sleeping like an Embryo

Sleeping bags in space are not as heavy as the earthbound models. In fact, they are rather thin. Their primary purpose is to keep the body from floating away and eventually bumping into the walls.

---

221. Brandon W. Maryatt, "Improvements to On-Orbit Sleeping Accommodations" (49th International Conference on Environmental Systems, Boston, MA, 2019), https://ntrs.nasa.gov/citations/20190027189.
222. James L. Broyan, Scott M. Cady, and David A. Welsh, "International Space Station Crew Quarters Ventilation and Acoustic Design Implementation" (International Conference on Environmental Systems, Barcelona, 2010), https://ntrs.nasa.gov/citations/20100017014.

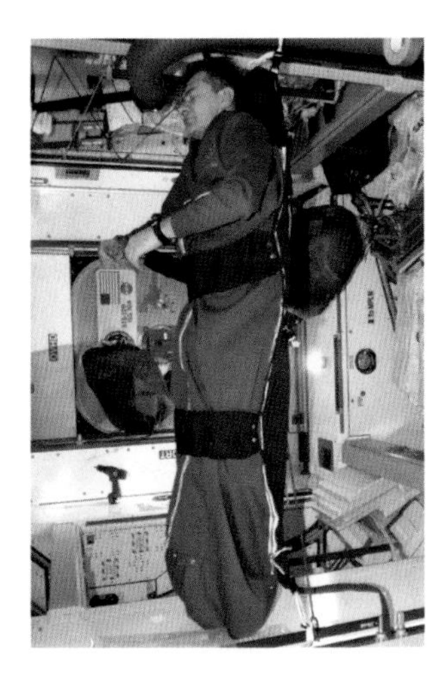
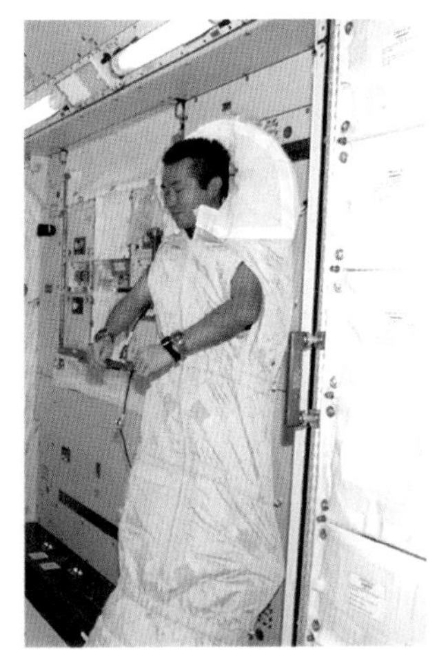

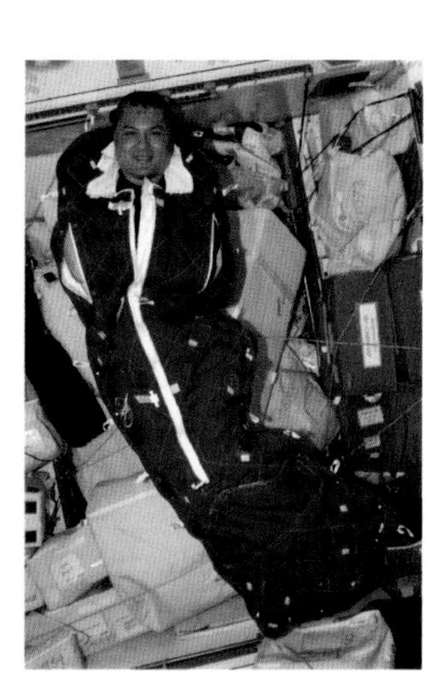

Evolution of astronaut sleeping bags. Left to right: 2007, 2009, 2014, 2022. Source: NASA.

Here are some design options for future space homesteads:

- **Option 1, Sleeping bag + vacuum pants:** As mentioned earlier, the human body naturally floats into a fetal position, with arms and legs drifting up from the body. A new sleeping bag would need to be accommodated for this. Below the waist, the vacuum pants would be incorporated into the sleeping bag, with as much noise reduction as possible. This device needs to be easy to float into and out of, easy to activate and allow for the legs to bend forward into a natural rest pose. (Most likely something simpler like MIT's Gravity Loading Countermeasure Skinsuit may be a better option.) The upper torso of the bag needs to have arm slits for access. A U-shaped headrest made of foam can be integrated into the top of the sleeping bag to gently keep the head turning at a minimum.
- **Option 2, Larger Crew Quarters:** The sleeping quarters need to be larger and heavily insulated from external noise. There should be quarters of various sizes for couples or even families. Privacy is a high priority, so these multi-person quarters would be rare exceptions. A continuous band of air intake/outtake vents (with temperature controls) should wrap around the top, sides, and bottom of the quarters. There needs to be a "desk," a personal area for art and memorabilia. Of course, there needs to be a window. It may not necessarily be a window to space, but it could have a view of the magnificent garden surrounding the living quarters.
- **Option 3: Sleeping Ring:** As initially described in Chapter 1, a modest-sized 30 meter (32.8 feet) diameter sleeping ring, based upon the NASA Nautilus-X ISS prototype, would allow your body to recover from a day in microgravity. With the inherent danger to the human body as described above, it makes more and more sense to develop some kind of crew quarters that can spin and generate some semblance of gravity. You've seen it in some science fiction films: while most of the space station would be microgravity, there can be areas that would be spinning to improve the comfort of the crew when they are in their personal quarters.

Conclusion

You made it this far. Congratulations!

I hope this book makes you realize that besides building better and safer rockets and space stations, there are many other human factors that need to be addressed if the human race wants to begin migrating out to the solar system. There are so many unknowns out there that we are only just beginning to understand, and with innovations we can begin to address them.

We need challenges. It helps us grow. One of the lesser known benefits of NASA is its Spinoffs program. Since 1976, over 2000 inventions and creative solutions derived from space projects have entered the commercial arena in the form of commercial goods, environmental cleanup, medicines, industrial processes, public safety, transportation, weather prediction, and information technology. That is just the tip of the iceberg.

One of my slogans is "Design for Space, Build for Earth." Each and every concept discussed in this book has the potential to reinvent and make better the originals here on Earth. From energy efficient washing machines, to improving the infrastructure for managing wastewater, hopefully this book will make you pause and rethink what the heck we are currently doing on Earth. We are creatures of habit: we do not like to change the way things are. Sometimes it takes a crisis, like a fire or global pandemic, to make people try something new. Other times, it takes a slick marketing campaign. Case in point: Steve Jobs stunned the world when he announced the Apple iPhone in 2007. This device's innovative touch screen interface merged so many abilities into one handheld object. Today, we all have one or more of these magic boxes in our possession. Will there ever be an invention like that to attract people to space?

Space travel has its allure. People want to go, but many things need to be developed to make it safer. I've done dozens of presentations on space infrastructure, the things that need to be built in order for thousands of humans to live off-world. One analogy is how cities get developed. You start with the pioneers who build the farms and small buildings. Then come the roads, then more buildings, and then other support systems. Below is a visual of some of the things that we have to build:

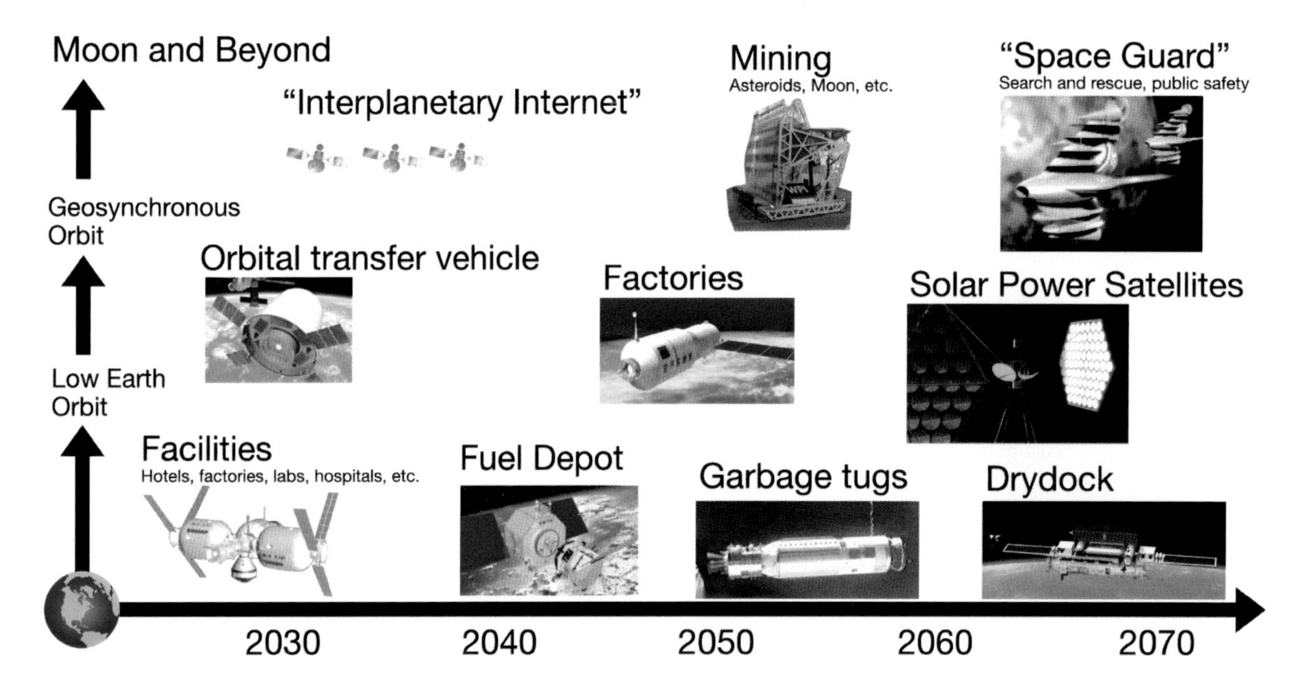

Orbital Infrastructure. Note: the dates are subject to change. Source: The author.

This list is not new. Many engineers and space experts have been talking for decades about the need for this infrastructure. But who is going to build it? How does a business make a profit on these critical items? Maybe some items just need to be a government service, that has mutual benefit for all? Either way, we are currently in a conundrum: we want to go to space, yet there is a lot of infrastructure that needs to be in place to make it safer for us to go. At the same time, it is hard for investors to justify spending money on space infrastructure projects until there are actual people out there.

While waiting for these big things to get built, I want to focus on the little things in space, the things that we can do today. Like enjoying a cup of tea while doing my laundry.

Acknowledgements

Holy cow! This one may take a while. There are so many people, living and dead, who have helped guide me and inspired me on this journey.

First of all, I would love to acknowledge my wife, Katherine Becvar, whose steadfast support and understanding of my wacky life is only rivaled by her own wacky life as a Librarian by day, and a costume designer by night.

To my son, Sebastian Apollo Coniglio, who is pretty much a Mini-Me, who constantly reminds me to have fun, and stay curious about life.

Many thanks to John Spencer of the Space Tourism Society. Years ahead of his time, John's ideas about Orbital Super Yachts, the Space Experience Economy and Earthbound space simulations inspired me to explore a different side of the space travel experience. Also, thanks to Allison Dollar for "herding the cats" each year and making the Space Tourism Conference a wonderful unique event! (https://spacetourismsociety.org/) and (https://www.spacetourismconf.com/).

To my friends and associates at the Space Frontier Foundation and the National Space Society, of whom there are too many people to count. Thank you very much for giving me insight and perspective to the space industry as a whole. (https://spacefrontier.org/) and (https://space.nss.org/). Also thanks to the Space Renaissance International, who during the pandemic gave me several opportunities to speak about my ideas via ZOOM calls. (https://agora.spacerenaissance.space/).

To Rick Tumlinson, the high priest of the NewSpace Movement, whose firebrand speeches charged up many of us Space Activists to keep trying to open that door to the High Frontier (https://ricktumlinson.com/). Also, thank you for writing the Foreword of the book!

Special thanks goes out to the Overview Roundtable! Created during the height of the Pandemic by "Overview Effect" author Frank White, with the help of Jared Angaza, Inara Tabir, and many others. This weekly ZOOM meeting of artists, astronomers, designers, educators, philosophers, engineers, entrepreneurs and more all have one thing in common: a love of space travel and spreading a positive vision of the future of the human race to everyone on Earth (https://www.humanspaceprogram.org/).

To the dynamic duo: George & Loretta Whitesides! Thank you two for being heroes and role models for the space activist community! George's quiet, reserved demeanor disguised a passion to help open the doors for commercial space inside NASA, and later as President of Virgin Galactic. Their enthusiasm spread the excitement about space via the Yuri's Night World Space Party, which is both a space education event packaged as a celebration (https://yurisnight.net/). Finally, Loretta's SpaceKind project is helping to make better people using space as the guiding force (https://www.spacekind.org/).

Much thanks to my friend Mikhail Baskov at iDare Space Travel for supporting me on this book adventure. (https://idare.space/)

Thanks to writing consultant Simon Golden for helping me stay focused and create a structure for the book. I've had many variations of this book in mind over the years, and with Simon‘s help I was able to streamline it and keep it super focused (https://www.simon-golden.com/).

ACTION GIRL! Andie Grace is a veteran Burner, and expert in self publishing and media relations, who helped me launch this book into orbit. Thanks for diving into the space world with me and guiding me through the complicated mess that is called book self-publishing.

Eternal gratitude to S.N. Jacobson for designing the book cover, and Sarah Dungan for doing her graphic design magic on the book cover image.

Thanks to Brent Heyning and Nick Donaldson for being awesome! Nick's Jedi Mastery of CAD software helped us design a unique cocktail glass that could be 3-D printed. And Brent's Hollywood prop making expertise helped us figure out many production details. (http://www.gotrobots.com/) (https://toyshoppesystems.com/)

Thank you Tim Bailey for advising us on the Zero Gravity Cocktail Glass Project, and giving us tips on understanding the effects of weightlessness on things. (https://novabailey.com/)

Thanks to Dr. Mark Weislogel, co-inventor of the Zero Gravity Coffee cup (with astronaut Donald Petit), for taking time to critique our Zero Gravity Cocktail Glass concept and for educating me on the power of blob management and microgravity plumbing. (https://www.pdx.edu/mechanical-materials-engineering/mark-weislogel)

Special thanks to Chris Sembroski, of Inspiration 4 fame, for letting me interview him and learning more details about flying inside a Dragon2 space capsule. Now I know the secret name of the Dragon2 space toilet! (https://inspiration4.com/).

Thanks to game designer and early private space traveler Richard Garriott, who believed in our crazy idea of designing and marketing a Zero Gravity Cocktail Glass, and gave me intimate details about what really happens in space when the cameras are turned off (https://richardgarriott.com/).

Thanks to my friends Gail Simpson and David Roberts, lending me moral and editorial support, and letting me use their secret hideaway to write. I was able to accelerate much progress in the book.

To my Mustachioed, Tweed Wearing Gentlemen: Paul Fuller, Evan Kinswood, and Andrew Crockett! Thanks for bouncing ideas and giving me advice as the project was going along. Also, thanks to Andew's partner-in-crime, Jane Beckman, for being the final editor at the very end!

Thanks to ESA Astronaut Samantha Christoforetti, CSA Astronaut Chris Hadfield, and NASA Astronaut Donald Petit: your research on improving the human experience in space (and excellent educational videos!) has inspired me and hopefully many future designers on creating better off-world living. I hope one day I can meet you all in person and thank you.

Thanks to Gene Forrer for giving me insights on the effects of radiation on the human body and how to prevent bad things from happening to you.

Thanks to Richard and Robert Godwin, fellow space nuts and publishers of the classic Apogee Books. Thanks for your insight on the publishing world (https://www.cgpublishing.com/).

Thanks to Carol Pinchefsky and Rebecca Donaldson for their advice on book writing and book publishing.

Thank you Robert Jacobson for giving me feedback and teaching me how to get your space book published (https://www.robertjacobson.com/book).

Much thanks to Misuzu Onuki, aerospace business consultant from Japan, and fellow space culture promoter! We participated in so many conferences and events together promoting a different side to living in space that most rocket geeks ignored. For years, we were the only voices discussing comfort, fashion, food, fun, and being human in an otherwise deadly environment at these space conferences. So glad that has finally changed!

Thanks to Izzy House for insights on marketing my book and for interviewing me for your podcast (http://izzy.house/).

Kudos to Chris Carberry for daring to write a book about alcohol in space, and not being worried about shaming by the more conservative members of the aerospace establishment (https://mcfarlandbooks.com/product/alcohol-in-space/).

Thanks to Grant Anderson of PARAGON Space Development Corporation, for helping me understand how NASA's life support system works (https://www.paragonsdc.com/).

Thanks to Rhonda Stevenson and Tim Alatorre from ABOVE: Space Development Corporation for your perspective on variable gravity space stations (https://abovespace.com/pioneer).

Thank you Dr. Marc Cohen for our discussions about space architecture and design. Thanks also for your feedback. (Yes, I now have an intro to artificial gravity, and added all your links and references.) I was there with him at the 2002 AIAA Space Architecture Symposium, and was a signatory for the Millenium Charter for the Space Architecture Mission Statement. Space Architecture as a field started very slowly and 20 years later it is now becoming a serious endeavor beyond just tiny groups inside NASA and ESA. I hope that one day I learn enough to be taken seriously. (http://spacearchitect.org/).

Thanks to my new ally, Phnam Bagley, and her crew at NONFICTION Design, whom I met as I was finishing up this book. I hope you continue to build more prototypes of space kitchen hardware and get to fly them in future space stations! (https://www.nonfiction.design/).

Thanks to Dr. Mark Nelson, one of the original Biosphere 2 adventurers, who introduced me to the concept of wastewater gardens to more efficiently work with nature to process human waste (https://ecotechnics.edu/).

I am grateful to the amazing books published by Lloyd Khan. Shelter Publications has an amazing collection of photo-filled books about simple shelters and off-the-grid living designed and built by regular folks, not professional architects. Their innovative ideas inspired me to use the homestead idea as a viable analogy for the next phase of long term living in space (https://www.shelterpub.com/).

Thanks to Vallejo Flood and Wastewater District, for giving me a tour of how wastewater is processed in an Earthbound situation. I am still confounded by how they go through all that effort to make the water 90% safe to drink, and yet cannot finish the cycle to 100%. I learned that this is a common problem throughout the world.

Thanks to programming wizard Kiki Jewell, designer Megan Lush, and my awesome wife Katherine Becvar for helping me create the TIKITRON Drinkbot. Also thanks to engineers Ken Mochel, Andrew Autore and fabricator Joe Phillips for helping me with the original COSMOBOT drinkbot. These Makers gave me insight on how fluids are managed and how to create an experience that is fun for people to enjoy! Speaking of which, thank you Dale Dougherty for creating MAKER Faire, which is an awesome showcase of DIY creativity and STEM/STEAM education! (https://makerfaire.com/). Also thanks to David Calkins and Simone Davalos for running Robogames, which is a true Olympics of hobby robotics! (http://robogames.net/index.php).

Much gratitude to Obtainium Works! Led by legendary artist and fabricator Shannon O'Hare and Cat Herder Extraordinaire (literally and metaphorically) Kathy O'Hare, this unique Steampunk kinetic arts collective taught me to take chances and learn to build and fix things with the stuff you got (called Obtainium) and build amazing art cars and other contraptions (https://www.obtainiumworks.net/).

Thanks to Barbara Brownie for publishing Space Wear, so I don't have to write a book on space fashion! (https://www.bloomsbury.com/us/spacewear-9781350000322/)

Thanks to Michael Rudis for his space sleeping bag concept.

Thanks to Elaine Swanson for teaching me which soaps are safe with plants.

Thanks to Daniel Thompkins for giving my insight on growing plants in microgravity.

Finally, a big group of thanks to these folks who took time to edit the book, give insightful comments, or gave me ideas: Dennis Wingo, Joy Montgomery, Babalou, David Armes, Kathryn Javandel, Nastia Illnichna, Doug Jones, Carla Uyeda, Aaron Kremmer, Carolyn M Jones.

My apologies if I missed anyone. Ping me and I will add you to the second edition!

Bibliography

8.01x—Lect 5—Circular Motion, Centripetal Forces, Perceived Gravity. Lectures by Walter Lewin. YouTube Channel. Massachusetts Institute of Technology (MIT), 2015. https://www.youtube.com/watch?v=mWj1ZEQTI8I.

"100km Altitude Boundary for Astronautics | World Air Sports Federation," August 1, 2017. https://www.fai.org/page/icare-boundary.

"A Citizen's Guide to Incineration." *United States Environmental Protection Agency* EPA 542-F-12-010 (September 2012): 2 pp. https://www.epa.gov/sites/default/files/2015-04/documents/a_citizens_guide_to_incineration.pdf.

ABOVE: Space Development Corporation. "ABOVE: Space Development Corporation." Accessed October 29, 2023. https://abovespace.com/.

Ackerman, Evan. "Autonomous Robots Are Helping Kill Coronavirus in Hospitals." IEEE Spectrum, March 11, 2020. https://spectrum.ieee.org/autonomous-robots-are-helping-kill-coronavirus-in-hospitals.

Adams, Constance M. "Define(Design)Ing the Human Domain: The Process of Architectural Integration in Long-Duration Space Facilities." SAE Technical Paper. Warrendale, PA: SAE International, July 13, 1998. https://doi.org/10.4271/981789.

Adams, Douglas. *Life, the Universe and Everything*. London: Pan Books Ltd., 1982.

"Advances in the Whipple Shield Design and Development: | Journal of Dynamic Behavior of Materials." Accessed November 7, 2023. https://link.springer.com/article/10.1007/s40870-021-00314-7.

Agency, Canadian Space. "Physical Activity in Space." Canadian Space Agency, August 18, 2006. https://www.asc-csa.gc.ca/eng/astronauts/living-in-space/physical-activity-in-space.asp.

———. "What Happens to Bones in Space?" Canadian Space Agency, August 18, 2006. https://www.asc-csa.gc.ca/eng/astronauts/space-medicine/bones.asp.

Alamalhodaei, Aria. "Gravitics Raises $20M to Make the Essential Units for Living and Working in Space." *TechCrunch* (blog), November 17, 2022. https://techcrunch.com/2022/11/17/gravitics-space-stations/.

Allen, Nafeesah, and Lowe Saddler. "What Is A Bidet? How Does It Work?" Forbes Home, November 1, 2022. https://www.forbes.com/home-improvement/bathroom/what-is-a-bidet/.

Almeida, Andres. "How 11 Deaf Men Helped Shape NASA's Human Spaceflight Program." NASA, May 4, 2017. http://www.nasa.gov/feature/how-11-deaf-men-helped-shape-nasas-human-spaceflight-program.

American Chemical Society. "Space-Grown Lettuce Could Help Astronauts Avoid Bone Loss," March 22, 2022. https://www.acs.org/content/acs/en/pressroom/newsreleases/2022/march/space-grown-lettuce-could-help-astronauts-avoid-bone-loss.html.

AstroAccess. "AstroAccess." Accessed December 29, 2022. https://astroaccess.org/.

"AstroAccess Successfully Completes 1st Weightless Research Flight with Int'l Disabled Crew—SatNews." Accessed October 29, 2023. https://news.satnews.com/2022/12/16/astroaccess-successfully-completes-1st-weightless-research-flight-with-intl-disabled-crew/.

Astronaut Chris Hadfield and Chef Traci Des Jardins Make a Space Burrito. Adam Savage's Tested: YouTube, 2013. https://www.youtube.com/watch?v=f8-UKqGZ_hs.

Astronaut Demos Drinking Coffee in Space. CollectSpace. International Space Station, 2008. https://www.youtube.com/watch?v=pk7LcugO3zg.

Astronaut Does Cosmic Kids Yoga IN SPACE!, 2022. https://www.youtube.com/watch?v=xWPBTqW3ipI.

Astronaut Donald Pettit on the Evolution of the Zero G Coffee Cup. Death Wish Coffee Company: Fueled by Death Cast, 2018. https://www.youtube.com/watch?v=ugQlivUuuXk.

AUSTRIA METALL SYSTEMTECHNIK (AMST) GmbH. "Barany Chair—Aerospace Medicine—AMST." Accessed January 21, 2024. https://www.amst.co.at/aerospace-medicine/training-simulation-products/barany-chair/.

Axiom Space. "Axiom Space, Prada Join Forces on Tech, Design for NASA's Next-Gen Lunar Spacesuits," October 4, 2023. https://www.axiomspace.com/news/prada-axiom-suit.

Axiom Space. "Mumm Announces Collaboration with Axiom Space," September 21, 2022. https://www.axiomspace.com/news/mumm-collaboration.

Baird, Gord, and Ann Baird. *Essential Composting Toilets: A Guide to Options, Design, Installation, and Use*. 1st Printing. British Columbia, Canada: New Society Publishers, 2018. https://newsociety.com/books/e/essential-composting-toilets.

Bandurski, Katie. "19 Plant-Based Meat Brands Every Vegetarian Needs to Know." *Taste of Home* (blog), November 15, 2022. https://www.tasteofhome.com/collection/vegetarian-brand-names/.

Barger, Laura K., Erin E. Flynn-Evans, Alan Kubey, Lorcan Walsh, Joseph M. Ronda, Wei Wang, Kenneth P. Wright, and Charles A. Czeisler. "Prevalence of Sleep Deficiency and Hypnotic Use Among Astronauts Before, During and After Spaceflight: An Observational Study." *The Lancet. Neurology* 13, no. 9 (September 2014): 904–12. https://doi.org/10.1016/S1474-4422(14)70122-X.

Barry, Patrick L., and Tony Phillips. "Mixed Up in Space." NASA. Science @NASA Blog, August 7, 2001. https://web.archive.org/web/20090513111327/http://science.nasa.gov/headlines/y2001/ast07aug_1.htm.

BathSelect. "Oxygenated Water Saving Oxygenics Comb Shower Head High Pressure ABS Plastic Clean Hair Brush for Bathroom | Shower Head Comb | Head Comb Suite." Accessed March 21, 2023. https://www.bathselect.com/product-p/bs10420.htm.

Beckmann, Sarah. "Dava Newman Presents 3D Knit BioSuit™ at 2022 MARS Conference." MIT Media Lab, March 30, 2022. https://www.media.mit.edu/posts/dava-newman-presents-3d-knit-biosuit-at-mars-conference/.

Bellisle, Rachel. "Project Overview ‹ The Gravity Loading Countermeasure Skinsuit." MIT Media Lab. Accessed June 17, 2023. https://www.media.mit.edu/projects/gravity-loading-countermeasure-skinsuit/overview/.

Bernasconi, Marco, Meindert Versteeg, and Roland Zenger. "A Multi-Purpose Astronaut Shower for Long-Duration Microgravity Missions." Valencia (Spain): International Astronautical Congress, October 2, 2006. https://www.researchgate.net/publication/258317776_A_Multi-Purpose_Astronaut_Shower_for_Long-Duration_Mirogravity_Missions.

Bevelacqua, Joseph John, and Seyed Mohammad Javad Mortazavi. "Commentary Regarding 'On-Orbit Sleep Problems of Astronauts and Countermeasures.'" *Military Medical Research* 5, no. 1 (October 30, 2018): 38. https://doi.org/10.1186/s40779-018-0185-2.

BioServe Space Technologies, University of Colorado Boulder, College of Engineering and Applied Science. "SABL Is BioServe's Next-Generation Smart Incubator," October 22, 2018. https://www.colorado.edu/center/bioserve/spaceflight-hardware/sabl.

Black Rock Rangers. "Black Rock Rangers." Accessed December 18, 2022. https://rangers.burningman.org/.

Bluth, B. J., and Martha Helppie. "Soviet Space Stations as Analogs, Second Edition." Washington, D.C.: NASA Headquarters, August 1, 1986. https://ntrs.nasa.gov/citations/19870012563.

Borden, Matthew, and Adam Dale. "ENY344/IN1248: Managing Plant Pests with Soaps." University of Florida, IFAS Extension, July 21, 2019. https://edis.ifas.ufl.edu/publication/IN1248.

Bourland, C.T., and G.L. Vogt. *The Astronaut's Cookbook*. 1st ed. New York: Springer Science + Business Media, LLC., 2010. https://link.springer.com/book/10.1007/978-1-4419-0624-3.

Bradbury, Peta, Hanjie Wu, Jung Un Choi, Alan E. Rowan, Hongyu Zhang, Kate Poole, Jan Lauko, and Joshua Chou. "Modeling the Impact of Microgravity at the Cellular Level: Implications for Human Disease." *Frontiers in Cell and Developmental Biology* 8 (February 21, 2020). https://doi.org/10.3389/fcell.2020.00096.

Britannica, The Editors of Encyclopaedia, ed. "Challenger Disaster." In *Encyclopedia Britannica*, August 21, 2022. https://www.britannica.com/event/Challenger-disaster.

Bronner, Lisa. "Dilutions Cheat Sheet for Dr. Bronner's Pure-Castile Soap." Dr. Bronner's, February 5, 2023. https://info.drbronner.com/all-one-blog/2023/02/dilutions-cheat-sheet-dr-bronners-pure-castile-soap/.

Brownie, Barbara. *Spacewear: Weightlessness and the Final Frontier of Fashion*. 1st ed. London: Bloomsbury Visual Arts, 2019. https://doi.org/10.5040/9781350000353.

Broyan, James L., Scott M. Cady, and David A. Welsh. "International Space Station Crew Quarters Ventilation and Acoustic Design Implementation." Barcelona, 2010. https://ntrs.nasa.gov/citations/20100017014.

Buis, Alan. "Earth's Magnetosphere: Protecting Our Planet from Harmful Space Energy." *NASA: Global Climate Change: Vital Signs of the Planet* (blog), August 3, 2021. https://climate.nasa.gov/news/3105/earths-magnetosphere-protecting-our-planet-from-harmful-space-energy.

Burners Without Borders. "Burners Without Borders." Accessed December 18, 2022. https://www.burnerswithoutborders.org/.

Burning Man Project. "Burning Man—Welcome Home." Accessed November 7, 2022. https://burningman.org.

Burning Man Project. "Burning Man Survival Guide 2023." Accessed July 25, 2023. https://survival.burningman.org/.

Butler, Carol. "Robert E. Stevenson Oral History." NASA Johnson Space Center, May 13, 1999. https://historycollection.jsc.nasa.gov/JSCHistoryPortal/history/oral_histories/StevensonRE/StevensonRE_5-13-99.htm.

Cain, Fraser. "Moss Grows in a Spiral... in Space." *Universe Today* (blog), January 27, 2005. https://www.universetoday.com/10228/moss-grows-in-a-spiral-in-space/.

Can The Human Body Handle Rotating Artificial Gravity? Scott Manley YouTube Channel, 2021. https://www.youtube.com/watch?v=nxeMoaxUpWk.

Canadian Space Agency. "Personal Hygiene in Space," August 18, 2006. https://www.asc-csa.gc.ca/eng/astronauts/living-in-space/personal-hygiene-in-space.asp.

Canright, Shelley. "The Physics of Space Gardens." NASA Education. Brian Dunbar, June 14, 2003. https://www.nasa.gov/audience/forstudents/5-8/features/space_gardens_feature.html.

Capri Sun Great Britain. "Our History | Learn More about the Capri-Sun Story | Capri-Sun UK." Accessed January 14, 2023. https://www.capri-sun.com/uk/about-us/history/.

Carberry, Chris. *Alcohol in Space.* 1st ed. McFarland & Company, 2019. https://mcfarlandbooks.com/product/alcohol-in-space/.

Carrington, Damian. "World's Largest Vats for Growing 'No-Kill' Meat to Be Built in US." *The Guardian*, May 25, 2022. https://www.theguardian.com/environment/2022/may/25/worlds-largest-vats-for-growing-no-kill-meat-to-be-built-in-us.

Centers for Disease Control and Prevention. "UV Radiation," July 5, 2022. https://www.cdc.gov/nceh/features/uv-radiation-safety/index.html.

"Challenger Disaster | Summary, Date, Cause, & Facts | Britannica." Accessed December 18, 2022. https://www.britannica.com/event/Challenger-disaster.

"Charles John Joughin : Titanic Survivor." Accessed January 21, 2024. https://www.encyclopedia-titanica.org/titanic-survivor/charles-john-joughin.html.

"Charles Joughin." In *Wikipedia*, July 7, 2023. https://en.wikipedia.org/w/index.php?title=Charles_Joughin&oldid=1164046144.

Chatters, Edward P., Brian J. Crothers, Air Command and Staff College, and Space Research Electives Seminars. "Space Surveillance Network." AU-18 Space Primer. Air University Press, 2009. https://www.jstor.org/stable/resrep13939.26.

Chris Hadfield—Nail Clipping in Space. Canadian Space Agency YouTube Channel. International Space Station, 2013. https://www.youtube.com/watch?v=xICkLB3vAeU.

Christiansen, Eric, Dana Lear, and Jim Hyde. "Micro-Meteoroid and Orbital Debris (MMOD) Protection Overview." NASA Johnson Space Center, October 17, 2018. https://ntrs.nasa.gov/citations/20190001193.

Ciotola, Mark. "Lettuce Garden Sent to ISS." SustainSpace, April 16, 2014. https://www.sustainspace.com/?tag=orbitec.

Cleveland Clinic. "Body Odor: Causes, Changes, Underlying Diseases & Treatment," March 4, 2022. https://my.clevelandclinic.org/health/symptoms/17865-body-odor.

Cocktail Robotics Grand Challenge, 2016. DNA Lounge, 2016. https://www.youtube.com/watch?v=Gcn8J7nHX3k.

Cohen, Marc. "Mockups 101: Code and Standard Research for Space Habitat Analogues." In *AIAA SPACE 2012 Conference & Exposition*, 34 pp. AIAA SPACE Forum. Pasadena, California: American Institute of Aeronautics and Astronautics, 2012. https://doi.org/10.2514/6.2012-5153.

Cohen, Marc, and Paul Houk. "Framework for a Crew Productivity Figure of Merit for Human Exploration." In *AIAA SPACE 2010 Conference & Exposition*, 29 pp. AIAA SPACE Forum. American Institute of Aeronautics and Astronautics, 2010. https://doi.org/10.2514/6.2010-8846.

Cohen, Marc M. "Comparative Configurations for Lunar Lander Habitation Volumes: 2005-2008." In *SAE International Journal of Aerospace*, 4:25 pp. Savannah, Georgia, USA, 2009. https://doi.org/10.4271/2009-01-2366.

———. "First Mars Habitat Architecture." In *AIAA SPACE 2015 Conference and Exposition*. American Institute of Aeronautics and Astronautics. Accessed January 21, 2024. https://doi.org/10.2514/6.2015-4517.

———. "Testing the Celentano Curve: An Empirical Survey of Predictions for Human Spacecraft Pressurized Volume." In *SAE International Journal of Aerospace*, 1:38 pp. San Francisco, California USA, 2008. https://doi.org/10.4271/2008-01-2027.

Collins Aerospace. "Crewed Missions," 2022. http://www.collinsaerospace.com/what-we-do/industries/space/crewed-missions.

Columbia University Irving Medical Center. "New Type of Ultraviolet Light Makes Indoor Air as Safe as Outdoors," March 17, 2022. https://www.cuimc.columbia.edu/news/new-type-ultraviolet-light-makes-indoor-air-safe-outdoors.

Coniglio, Samuel. "DRINKBOTS." Obtainium Works. Accessed November 24, 2022. https://www.obtainiumworks.net/drinkbots.

Connors, M. M., A. A. Harrison, and F. R. Akins. "Living Aloft: Human Requirements for Extended Spaceflight." NASA Ames Research Center, January 1, 1985. https://ntrs.nasa.gov/citations/19850024459.

Constantin, Margaux, Steve Saxon, and Jackey Yu. "Reimagining the $9 Trillion Tourism Economy—What Will It Take?" McKinsey & Company, August 5, 2020. https://www.mckinsey.com/industries/travel-logistics-and-infrastructure/our-insights/reimagining-the-9-trillion-tourism-economy-what-will-it-take.

Continuum of Space Architecture | Marc M. Cohen. Emerging Fields in Architecture. Vienna, Austria: Technische Universität Wien, 2023. https://www.youtube.com/watch?v=aSjOYHxiS_s.

Daley, Jason. "With a 'Zero G' Oven, Astronauts Can Have Their Cookies, but They Can't Eat Them Too." Smithsonian Magazine, November 5, 2019. https://www.smithsonianmag.com/smart-news/space-no-one-can-hear-you-nom-space-station-getting-cookie-baking-oven-1-180973455/.

David, Leonard. "Having 'Skinsuit' in the Game: Managing Microgravity." *Leonard David's INSIDE OUTER SPACE* (blog), May 20, 2023. https://www.LeonardDavid.com/having-skinsuit-in-the-game-managing-microgravity/.

De La Fuente, Horacio, Jasen L. Raboin, Gary R. Spexarth, and Gerard D. Valle. "TransHab: NASA's Large-Scale Inflatable Spacecraft," 9. Atlanta, GA, 2000. https://ntrs.nasa.gov/citations/20100042636.

Delany, Alex. "So, What Is Sous Vide, Anyway?" *Bon Appétit: Cooking* (blog), January 24, 2018. https://www.bonappetit.com/story/what-is-sous-vide-cooking.

Drehobl, Marissa, Matthew Herman, and Victoria Rduch. "Strigil: Boudoir, Bath and Temple." Amherst College, Mead Art Museum, 2009. https://www.amherst.edu/museums/mead/resources_faculty/faculty/courseproj/boudoir/strigil.

Ebrahimnejad, Ramin. "Lab-Grown Fruits." *Association for Vertical Farming* (blog), January 15, 2021. https://vertical-farming.net/blog/2021/01/15/fruit-of-knowledge/.

Editor, SpaceRef. "Space Entertainment Enterprise (S.E.E.) Announces World's First Entertainment Arena and Content Studios in Space, Built by Axiom Space." SpaceRef, January 20, 2022. https://spaceref.com/press-release/space-entertainment-enterprise-see-announces-worldaeurs-first-entertainment-arena-and-content-studios-in-space-built-by-axiom-space/.

Edwards, Abigail. "Can Humans Survive in Space Without a Space Suit?" *Penn State University, SiOWfa16: Science in Our World: Certainty and Controversy* (blog), September 11, 2016. https://sites.psu.edu/siowfa16/2016/09/11/can-humans-survive-in-space-without-a-spacesuit/.

Elberfeld, Alice, Cody Beard, and Beth Westfall. "Artificial Gravity Research." NASA Langley Research Center, July 31, 2020. https://ntrs.nasa.gov/citations/20205005149.

Electrolux Professional Global. "Wet Cleaning vs Dry Cleaning | Electrolux Professional," June 21, 2023. https://www.electroluxprofessional.com/wet-cleaning-vs-dry-cleaning-pros-and-cons/.

Ellis, Justin, Jonathan Bigelow, Connor Shelander, Dennis Chertkovsky, Michael Ewert, Melissa McKinley, and Ayyoub Momen. "Ultrasonic Clothes Washer/Dryer Combination for Moon, Mars and ISS Applications." Calgary, Alberta: NASA Johnson Space Center. Accessed January 18, 2024. https://ntrs.nasa.gov/citations/20230001711.

Espiritu, Kevin. "The Benefits of Moss in Your Garden: Add Some Character to Your Garden." *PartSelect.Com* (blog), August 9, 2017. https://www.partselect.com/blog/grow-moss-in-your-garden/.

Essentra Components. "UV and Its Effect on Plastics: An Overview," January 23, 2019. https://www.essentracomponents.com/en-us/news/manufacturing/injection-molding/uv-and-its-effect-on-plastics-an-overview.

Evers, Jeannie. "Meteor." National Geographic Society, October 19, 2023. https://education.nationalgeographic.org/resource/meteor.

FAZIO, G. G., J. V. JELLEY, and W. N. CHARMAN. "Generation of Cherenkov Light Flashes by Cosmic Radiation within the Eyes of the Apollo Astronauts." *Nature* 228, no. 5268 (October 1, 1970): 260–64. https://doi.org/10.1038/228260a0.

Ferdowsi, Samir. "How Far Away Are We From Downloading Our Clothes?" Refinery 29, April 29, 2021. https://www.refinery29.com/en-us/3d-printing-fashion.

Fitless Humans (WALL·E). Pixar Animation Studio, 2013. https://www.youtube.com/watch?v=s-kdRdzxdZQ.

Flynn, Michael, Reneé Matossian, Marc Cohen, Sherwin Gormly, and Rocco Mancinelli. "Membrane Based Habitat Wall Architectures for Life Support and Evolving Structures." Barcelona, Spain: American Institute of Aeronautics and Astronautics, 2012. https://doi.org/10.2514/6.2010-6073.

Flynn, Michael T., Marc Cohen, Renee L. Matossian, Sherwin Gormly, Rocco Mancinelli, Jack Miller, Jurek Parodi, and Elyssee Grossi. "Water Walls Architecture: Massively Redundant And Highly Reliable Life Support For Long Duration Exploration Missions," November 12, 2018. https://ntrs.nasa.gov/citations/20190001191.

Foust, Jeff. "Bigelow Aerospace Transfers BEAM Space Station Module to NASA." SpaceNews, January 21, 2022. https://spacenews.com/bigelow-aerospace-transfers-beam-space-station-module-to-nasa/.

———. "Decadal Survey Recommends Massive Funding Increase for NASA Biological and Physical Sciences." *SpaceNews* (blog), September 13, 2023. https://spacenews.com/decadal-survey-recommends-massive-funding-increase-for-nasa-biological-and-physical-sciences/.

———. "The Space Review: Shaking up the Commercial Space Station Industry." The Space Review, October 30, 2023. https://www.thespacereview.com/article/4681/1.

Friedman, Virginia, Melissa Wagner, and Nancy Armstrong. *Stains: A Spotter's Guide*. New York: Barnes & Noble Books, 2005.

From Idea To Infinity. Space Games Federation, 2019. https://vimeo.com/320018239.

Garcia, Mark. "Space Debris and Human Spacecraft." Text. NASA, April 13, 2015. http://www.nasa.gov/mission_pages/station/news/orbital_debris.html.

Garver, Lori. *Escaping Gravity: My Quest to Transform NASA and Launch a New Space Age*. First Edition. Diversion Books, 2022. https://diversionbooks.com/books/escaping-gravity/.

Genesis Engineering. "SPS." Accessed October 29, 2023. https://genesisesi.com/projects/sps/.

Granath, Bob. "Lunar, Martian Greenhouses Designed to Mimic Those on Earth." Text. NASA, April 24, 2017. http://www.nasa.gov/feature/lunar-martian-greenhouses-designed-to-mimic-those-on-earth.

"Gravitics." Accessed November 27, 2023. https://www.gravitics.com/.

Gravitics Inc. "Gravitics—Starmax." Accessed January 10, 2024. https://www.gravitics.com/starmax.

Gravity, Space and Architecture | Ted W. Hall (University of Michigan). Emerging Fields in Architecture. Vienna, Austria: Technische Universität Wien, 2022. https://www.youtube.com/watch?v=2KK-Mq8FL7w.

gravityLab. "gravityLab — Accelerating Life Off-Planet." Accessed May 10, 2023. https://www.gravitylabspace.com.

Green, J. A., and J. L. Peacock. "Effects of Simulated Artificial Gravity on Human Performance." NASA Langley Research Center, November 1, 1972. https://ntrs.nasa.gov/citations/19730003384.

Green, J. A., J. L. Peacock, and A. P. Holm. "A Study of Human Performance in a Rotating Environment." NASA Langley Research Center, January 1, 1971. https://ntrs.nasa.gov/citations/19720019454.

Greywater Action. "Choosing and Irrigating Plants with Greywater," March 1, 2023. https://greywateraction.org/greywater-choosing-plants-and-irrigating/.

Greywater Action. "Greywater Reuse," March 1, 2023. https://greywateraction.org/greywater-reuse/.

Griffin, Brand. "Benefits of a Single-Person Spacecraft for Weightless Operations." In *42nd International Conference on Environmental Systems*. International Conference on Environmental Systems (ICES). American Institute of Aeronautics and Astronautics, 2012. https://doi.org/10.2514/6.2012-3630.

Griffin, Brand N., Robert Rashford, Samuel Gaylin, Dylan Bell, and John Harro. "The Wait-Less EVA Solution: Single-Person Spacecraft." ASCEND. Virtual Event: American Institute of Aeronautics and Astronautics, 2020. https://doi.org/10.2514/6.2020-4170.

Griffin, Brand N., Robert Rashford, Christopher Tolman, and Samuel Gaylin. "Single-Person Spacecraft Provides Commercially Viable EVA Including Tourist Excursions for Orbital Reef," 8 pages. Las Vegas, Nevada: American Institute of Aeronautics and Astronautics, 2022. https://doi.org/10.2514/6.2022-4209.

Griffin, Brand Norman, Robert Howard, Sudhakar Rajulu, and David Smitherman. "Creating a Lunar EVA Work Envelope." Savannah, Georgia, USA: Society of Automotive Engineers, 2009. https://spacearchitect.org/pubs/SAE-2009-01-2569.pdf.

Groh, Jamie. "New Space Race Defines Where to Live and Work in Orbit, No Longer about How to Get There." News. Florida Today, March 27, 2022. https://www.floridatoday.com/story/tech/science/space/2022/03/27/nasa-funds-four-ideas-commercial-space-stations-plans-iss-deorbit/6782521001/.

Hall, Loura. "Space Farming Yields a Crop of Benefits for Earth." Text. NASA, August 6, 2015. http://www.nasa.gov/feature/space-farming-yields-a-crop-of-benefits-for-earth.

Hall, Theodore. "Artificial Gravity Visualization, Empathy, and Design." In *Space 2006*, 22. AIAA SPACE Forum, AIAA-2006-7321. San Jose, CA: American Institute of Aeronautics and Astronautics, 2006. https://doi.org/10.2514/6.2006-7321.

Hall, Theodore, and Al Globus. "Space Settlement Population Rotation Tolerance." *NSS Space Settlement Journal* #2 (June 2017): 25 pp. https://doi.org/10.13140/RG.2.2.29467.57123.

Hall, Theodore W. "Artificial Gravity." Text. Theodore W. Hall, January 3, 2024. https://www.artificial-gravity.com/.

———. "Artificial Gravity in Theory and Practice," 20 pp. Vienna, Austria: 46th International Conference on Environmental Systems, 2016. http://hdl.handle.net/2346/67587.

———. "Gravity as an Environmental System," 2000-01–2244:12. Toulouse, France: SAE International, 2000. https://saemobilus.sae.org/content/2000-01-2244/.

———. "GRAVITY, SPACE, AND ARCHITECTURE." In *2001: Building for Space Travel*, edited by J. Zukowsky, 168–74. New York, New York, USA: Harry N. Abrams, Inc., 2001. http://spacearchitect.org/pubs/Hall-2001-GravitySpaceArch-prepub.pdf.

———. "Space Architecture Publications." Text. SpaceArchitect.org, January 14, 2024. https://spacearchitect.org/pubs/pub-biblio.htm.

———. "SpinCalc." Theodore W. Hall, September 9, 2018. http://www.artificial-gravity.com/sw/SpinCalc/.

———. "The Architecture of Artificial Gravity Habitats." Presented at the Future in Space Operations (FISO) Colloquium, November 17, 2010. https://web.archive.org/web/20140331025927/http://spirit.as.utexas.edu/~fiso/telecon/Hall_11-17-10/Hall_11-17-10.pdf.

Hall-Geisler, Kristen. "How Do Astronauts Shower in Space?" HowStuffWorks, January 22, 2021. https://science.howstuffworks.com/shower-in-space.htm.

Handcrafted Soap & Cosmetic Guild. "Types of Handcrafted Soap." Accessed December 7, 2023. https://www.soapguild.org/buy-handcrafted/benefits/types-of-soap.

"Haven-1 Commercial Space Station + Haven-1 Mission Revealed by Vast along with SpaceX Involvement—SatNews." Accessed October 29, 2023. https://news.satnews.com/2023/05/10/haven-1-commercial-space-station-haven-1-mission-revealed-by-vast-along-with-spacex-involvement/.

Heiney, Anna. "Growing Plants in Space." Text. NASA, Kennedy Space Center, July 12, 2021. http://www.nasa.gov/content/growing-plants-in-space.

Henderson, Edward M., and Mark L. Holderman. "Technology Applications That Support Space Exploration." San Diego, CA: AIAA, 2011. https://ntrs.nasa.gov/citations/20110013138.

Holderman, Mark L. "Nautilus-X: Multi-Mission Space Exploration Vehicle." NASA Technology Applications Assessment Team, FISO Telecon presentation, January 26, 2011. https://web.archive.org/web/20110304044259/http://spirit.as.utexas.edu/~fiso/telecon/Holderman-Henderson_1-26-11/Holderman_1-26-11.ppt.

House, Izzy. *Space Marketing: Competing in the New Commercial Space Industry*. 1st Edition. USA: Self-published, Izzy House, 2021. https://izzy.house/space-marketing-book/.

How to Prepare (Thanksgiving) Food in Space. NASA Johnson Space Center, 2015. https://www.youtube.com/watch?v=60fxGvNLFtY.

Howe, A. Scott, Brent Sherwoood, Damon Landau, and Theodore W. Hall. "Gateway Gravity Testbed (GGT)," ICES-2029-23:14. Boston, Massachusetts, 2019. https://ttu-ir.tdl.org/bitstream/handle/2346/84708/ICES-2019-23.pdf.

Howell, Elizabeth. "New SpaceX Spacesuits Get Five-Star Rating from NASA Astronauts." Space.com, June 10, 2020. https://www.space.com/spacex-spacesuits-five-star-astronaut-review.html.

Howell, Elizabeth. "Showering in Space: Astronaut Home Video Shows Off 'Hygiene Corner.'" Space.com, June 9, 2015. https://www.space.com/29610-showering-in-space-astronaut-video.html.

Hua, Jonathan. "The Mars Farm: A Not-Too-Distant Reality?" The Spoon, January 25, 2021. https://thespoon.tech/the-mars-farm-a-not-too-distant-reality/.

Hurlich, A. "Spaced Armor," 24. Aberdeen, Maryland: Watertown Arsenal Labs, 1950. https://apps.dtic.mil/sti/citations/ADA954865.

Hutchings, Kristy. "NASA Partners with Long Beach's Vast Space on Artificial Gravity Station Development." *Press Telegram* (blog), June 16, 2023. https://www.presstelegram.com/2023/06/16/nasa-partners-with-long-beachs-vast-space-on-artificial-gravity-station-development/.

Incineration, National Research Council (US) Committee on Health Effects of Waste. "Incineration Processes and Environmental Releases." In *Waste Incineration & Public Health*. National Academies Press (US), 2000. https://www.ncbi.nlm.nih.gov/books/NBK233627/.

Infinitia Research. "Hydrophobic Substances: What Are They and What Are They Used For?," May 26, 2021. https://www.infinitiaresearch.com/en/news/hydrophobic-substances-what-are-they-and-what-are-they-used-for/.

Irish, Perla. "Marsh Plants That Clean Greywater." *Housesumo.Com* (blog), September 25, 2019. https://www.housesumo.com/marsh-plants-that-clean-greywater/.

Jabr, Ferris. "Why Soap Works." *The New York Times*, March 13, 2020, sec. Health. https://www.nytimes.com/2020/03/13/health/soap-coronavirus-handwashing-germs.html.

Jane Poynter: Life in Biosphere 2, 2009. https://www.youtube.com/watch?v=a7B39MLVelc.

Jeanne Morel et Paul Marlier—Art In Space, 2021. https://www.youtube.com/watch?v=E178DbabTd8.

Jeff, and Dustin. "Fueled By Death Cast Ep. 106—DONALD PETTIT." Video. Death Wish Coffee Video Podcast: Fueled by Death Cast. Accessed November 24, 2022. https://www.deathwishcoffee.com/pages/fbdc-ep-106-donald-pettit.

Jenkins, Joseph. *The Humanure Handbook: Shit in a Nutshell*. 4th ed. Grove City, PA: Joseph Jenkins, Inc., 2019. https://slateroofwarehouse.com/Books/Joseph_Jenkins_Books/Humanure_Handbook.

Johnson, Richard D., and Charles Holbrow. "Space Settlements: A Design Study." NASA Ames Research Center, January 1, 1977. https://ntrs.nasa.gov/citations/19770014162.

Jones, Andrew, and Daisy Dobrijevic. "China's Space Station, Tiangong: A Complete Guide." Space.com, August 24, 2021. https://www.space.com/tiangong-space-station.

Junaedi, Christian, Kyle Hawley, Dennis Walsh, Subir Roychoudhury, Morgan B. Abney, and Jay L. Perry. "Compact and Lightweight Sabatier Reactor for Carbon Dioxide Reduction." Portland, OR, 2011. https://ntrs.nasa.gov/citations/20120016419.

Kahn, Lloyd. *Rolling Homes: Shelter on Wheels*. 1st Edition. Bolinas, CA: Shelter Publications, 2022. https://www.shelterpub.com/building/rolling-homes.

———. *THE SEPTIC SYSTEM OWNER'S MANUAL*. 2nd Edition. Bolinas, CA: Shelter Publications, 2007. https://www.shelterpub.com/building/the-septic-system-owners-manual.

———. *Tiny Homes: Simple Shelter*. 1st Edition. Bolinas, CA: Shelter Publications, 2012. https://www.shelterpub.com/building/tiny-homes.

Kahn, Lloyd, and Leslie Creed. *The Half-Acre Homestead: 46 Years of Building and Gardening*. 1st Edition. Bolinas, CA: Shelter Publications, 2020. https://www.shelterpub.com/building/halfacrehomestead.

Kahn, Lloyd, and Bob Easton. *Shelter 2*. 2nd Edition. Bolinas, CA: Shelter Publications, 2010. https://www.shelterpub.com/building/shelter-2.

Kanis, Simeon. "What is Astrobee?" Text. NASA, November 8, 2016. http://www.nasa.gov/astrobee.

Kim, Shi En. "The Quest to Build a Functional, Energy-Efficient Refrigerator That Works in Space." Smithsonian Magazine, July 27, 2021. https://www.smithsonianmag.com/innovation/quest-to-build-functional-energy-efficient-refrigerator-that-works-in-space-180978281/.

Kluger, Jeffrey. "The World's First Space Tourist Plans a Return Trip—This Time to the Moon." Time, October 17, 2022. https://time.com/6222212/dennis-tito-moon-space-tourism/.

Korpela, Jyrki. “What Is Surface Tension?” *Biolin Scientific* (blog), October 16, 2018. https://www.biolinscientific.com/blog/what-is-surface-tension.

Kowal, Mary Robinette. “The Need for Caffeine Was the Mother of Invention.” *The New York Times*, November 2, 2020, sec. Science. https://www.nytimes.com/2020/11/02/science/space-station-coffee.html.

Kulu, Erik. “Space Farming—Factories in Space.” Factories in Space, March 12, 2023. https://www.factoriesinspace.com/space-farming.

Lagomarsino, Valentina. “Hydroponics: The Power of Water to Grow Food.” *Science in the News* (blog), September 26, 2019. https://sitn.hms.harvard.edu/flash/2019/hydroponics-the-power-of-water-to-grow-food/.

Leasca, Stacey. “5 Things to Know Before Ordering Food and Drinks on the Plane.” Food & Wine, December 6, 2022. https://www.foodandwine.com/food-drink-airplane-ordering-tips-6836206.

Lee, Min-sun, Juyoung Lee, Bum-Jin Park, and Yoshifumi Miyazaki. “Interaction with Indoor Plants May Reduce Psychological and Physiological Stress by Suppressing Autonomic Nervous System Activity in Young Adults: A Randomized Crossover Study.” *Journal of Physiological Anthropology* 34, no. 1 (April 28, 2015): 21. https://doi.org/10.1186/s40101-015-0060-8.

“Lessons Learned on the Skylab Program.” Houston, TX: LYNDON B. JOHNSON SPACE CENTER, July 18, 1974. https://ntrs.nasa.gov/citations/19760004100.

Let’s Talk Science. “Temperature on Earth and on the ISS,” September 23, 2019. https://letstalkscience.ca/educational-resources/backgrounders/temperature-on-earth-and-on-iss.

LIBRARY, NASA/SCIENCE PHOTO. “Weightless Test of Shower for Space Station—Stock Image—S560/0189.” Science Photo Library. Accessed March 21, 2023. https://www.sciencephoto.com/media/337077/view/weightless-test-of-shower-for-space-station.

Lockley, Steven W., and George C. Brainard. “Lighting Effects.” NASA Technical Reports Server, June 7, 2016. https://ntrs.nasa.gov/citations/20160006727.

Loon, Jacco van. “Could the Earth Ever Stop Spinning, and What Would Happen If It Did?” Space.com, January 23, 2022. https://www.space.com/what-if-earth-stopped-spinning.

Maindl, Thomas I., Roman Miksch, and Birgit Loibnegger. “Stability of a Rotating Asteroid Housing a Space Station.” arXiv, December 26, 2018. https://doi.org/10.48550/arXiv.1812.10436.

Maker Faire. “Maker Faire Bay Area.” Accessed July 13, 2023. https://makerfaire.com/bay-area/.

Maker Faire. “Maker Faire: Explore. Play. Learn. Create. Invent.” Accessed December 18, 2022. https://makerfaire.com/.

Mandelbaum, Ryan F., and Rae Paoletta. “We Chatted With an Astronaut About Showering, Farting, and Boning in Space.” Gizmodo, April 21, 2017. https://gizmodo.com/we-chatted-with-an-astronaut-about-showering-farting-1794538749.

Marble, Organic. “Answer to ‘How Did Skylab’s “Space Shower” Work?’” *Space Exploration Stack Exchange*, May 11, 2019. https://space.stackexchange.com/a/36082.

———. “Was There Really a Shuttle Toilet Training Device with a ‘Boresight Camera’?” Forum post. *Space Exploration Stack Exchange*, June 25, 2020. https://space.stackexchange.com/q/40338.

Mark Nelson. *The Wastewater Gardener: Preserving the Planet One Flush at a Time*. 1st Edition. Santa Fe, NM: Synergetic Press, 2014. https://synergeticpress.com/catalog/the-wastewater-gardener/.

Mars, Kelli. "30 Years Ago: Daniel Goldin Sworn in as NASA's Ninth Administrator." Text. NASA History, March 31, 2022. http://www.nasa.gov/feature/30-years-ago-daniel-goldin-sworn-in-as-nasa-s-ninth-administrator.

Mars, Kelli, Amy Blanchett, and Laurie Abadie. "Space Radiation Is Risky Business for the Human Body." Text. *NASA* (blog), September 18, 2017. http://www.nasa.gov/feature/space-radiation-is-risky-business-for-the-human-body.

Maryatt, Brandon W. "Improvements to On-Orbit Sleeping Accommodations." Boston, MA, 2019. https://ntrs.nasa.gov/citations/20190027189.

Mathias, Jennifer. "What Is Outgassing Testing?" *Innovatech Labs* (blog), January 26, 2016. https://www.innovatechlabs.com/newsroom/882/outgassing-testing/.

May, Sandra. "Eating in Space." Text. NASA, June 8, 2015. http://www.nasa.gov/audience/foreducators/stem-on-station/ditl_eating.

———. "What Is Microgravity?" NASA. *What Is Microgravity?* (blog), February 15, 2012. https://www.nasa.gov/audience/forstudents/5-8/features/nasa-knows/what-is-microgravity-58.html.

McGee, Tim. "Ancient Masters of Healthy Skin." *Medium* (blog), May 20, 2015. https://medium.com/@plumeriatiki/ancient-masters-of-healthy-skin-b0e87d1511ed.

Mckenzie, Kevin Hinton & Ryan. "Tru Earth." Tru Earth. Accessed November 2, 2022. https://www.tru.earth.

McKinley, Melissa, James Lee Broyan, Jr., Laura Shaw, Donald L Carter, and Jim Fuller. "NASA Universal Waste Management System and Toilet Integration Hardware Delivery and Planned Operation on ISS." In *50th International Conference on Environmental Systems*, Vol. ICES-2021-403, 2021. https://ttu-ir.tdl.org/bitstream/handle/2346/87295/ICES-2021-403.pdf?sequence=1&isAllowed=n.

McVean, Ada. "The Head Baker of the Titanic Spent Two Hours in Frigid Water and Emerged with Only Swollen Feet!" *Office for Science and Society, McGill University* (blog), July 3, 2020. https://www.mcgill.ca/oss/article/history-did-you-know/head-baker-titanic-spent-two-hours-frigid-water-and-emerged-only-swollen-feet.

Memory Alpha. "Replicator," November 27, 2023. https://memory-alpha.fandom.com/wiki/Replicator.

Middleton, R. L., A. C. Krupnick, J. C. Reily, and B. J. Schrick. "Design, Development, and Operation of a Zero Gravity Shower," Vol. AAS PAPER 74-136. Los Angeles, CA, 1974. https://ntrs.nasa.gov/citations/19740059331.

Mohanty, Susmita. "Design Concepts for Zero-G Whole Body Cleansing on ISS Alpha—Part II: Individual Design Project," September 1, 2001. https://ntrs.nasa.gov/citations/20010098604.

Mohon, Lee. "Environmental Control and Life Support System (ECLSS)." Text. NASA, September 11, 2017. http://www.nasa.gov/centers/marshall/history/eclss.html.

Morel, Jeanne, and Paul Marlier. "ART IN SPACE." ART IN SPACE. Accessed October 27, 2022. https://www.artinspace.fr.

Murray, Daniel H., Andrew A. Pilmanis, Rebecca S. Blue, James M. Pattarini, Jennifer Law, C. Gresham Bayne, Matthew W. Turney, and Jonathan B. Clark. "Pathophysiology, Prevention, and Treatment of Ebullism." *Aviation, Space, and Environmental Medicine* 84, no. 2 (February 2013): 89–96. https://doi.org/10.3357/asem.3468.2013.

Museum of Decorative Arts and Design. "Octave de Gaulle, Civilizing Space," July 2, 2017. https://madd-bordeaux.fr/en/exhibitions/octave-de-gaulle-civilizing-space.

NASA Blog: A Lab Aloft (International Space Station Research). "Space Station Espresso Cups: Strong Coffee Yields Stronger Science—A Lab Aloft (International Space Station

Research).” Blog, May 1, 2015. https://blogs.nasa.gov/ISS_Science_Blog/2015/05/01/space-station-espresso-cups-strong-coffee-yields-stronger-science/.

NASA Goddard Space Flight Center. “Description | Outgassing.” Accessed October 23, 2022. https://outgassing.nasa.gov/Description.

NASA Image and Video Library. “ISSEspresso | NASA Image and Video Library,” May 3, 2015. https://images.nasa.gov/details-iss043e160068.

NASA John C. Stennis Space Center. “SSC’s Environmental Assurance Program.” Accessed March 21, 2023. https://www.ssc.nasa.gov/environmental/docforms/water_research/water_research.html.

NASA Science. “10 Things: What’s That Space Rock?,” July 21, 2022. https://science.nasa.gov/solar-system/10-things-whats-that-space-rock/.

NASA Space Station Research Explorer. “P&G Telescience Investigation of Detergent Experiments,” December 10, 2021. https://www.nasa.gov/mission_pages/station/research/experiments/explorer/Investigation.html?#id=8595.

NASA Technology Transfer Program. “Pre-Treatment Solution for Water Recovery.” Accessed October 25, 2022. https://technology.nasa.gov/patent/MSC-TOPS-68.

National Health Service UK. “Embolism,” October 18, 2017. https://www.nhs.uk/conditions/embolism/.

National Institute of Health (NIH), National Institute of General Medical Sciences (NIGMS). “Circadian Rhythms,” May 4, 2022. https://nigms.nih.gov/education/fact-sheets/Pages/circadian-rhythms.aspx.

National Research Council (US) Committee on Health Effects of Waste Incineration. *Waste Incineration & Public Health*. Washington (DC): National Academies Press (US), 2000. http://www.ncbi.nlm.nih.gov/books/NBK233629/.

Nelson, Mark. “Projects—Biosphere 2—The Institute of Ecotechnics.” *Institute of Ecotechnics* (blog). Accessed October 26, 2022. https://ecotechnics.edu/projects/biosphere-2/.

———. “The Challenge of Managing Water and Nutrient Cycles in a Mini-World—the Lessons from Biosphere 2.” *Global Water Forum* (blog), August 6, 2020. https://globalwaterforum.org/2020/08/06/the-challenge-of-managing-water-and-nutrient-cycles-in-a-mini-world-the-lessons-from-biosphere-2/.

Neufeld, Michael. “Mars Project: Wernher von Braun as a Science-Fiction Writer.” *National Air & Space Museum, Smithsonian Institution* (blog), January 22, 2021. https://airandspace.si.edu/stories/editorial/mars-project-wernher-von-braun-science-fiction-writer.

Nevills, Amiko. “NASA—Food in Space Gallery.” NASA Johnson Space Center, November 25, 2007. https://www.nasa.gov/audience/formedia/presskits/spacefood/gallery_iss005e16310.html.

———. “NASA—Space Food Laboratory Gallery.” NASA Johnson Space Center, November 25, 2007. https://www.nasa.gov/audience/formedia/presskits/spacefood/gallery_jsc2003e63872.html.

Nixon, David A. “Technology Demonstrator for a Rotating Space Station.” *Journal of the British Interplanetary Society* 75 (June 6, 2022): 209–17. https://www.artificial-gravity.com/JBIS-75-Nixon.pdf.

OIKOFUGE. “Coriolis Effect In A Rotating Space Habitat.” The Oikofuge, March 27, 2017. https://oikofuge.com/coriolis-effect-rotating-space-habitat/.

OK Go—Upside Down & Inside Out, 2016. https://www.youtube.com/watch?v=LWGJA9i18Co.

OK Go—Upside Down & Inside Out BTS—How We Did It, 2016. https://www.youtube.com/watch?v=pnTqZ68fI7Q.

O’Neill, Gerard K. *The High Frontier: Human Colonies in Space*. 3rd Edition. Burlington, Ontario, Canada: Apogee Books, 2000. https://www.cgpublishing.com/Books/Highfrontier.html.

Orbital Assembly. “Orbital Assembly.” Accessed December 11, 2022. https://orbitalassembly.com/.

Pai, A., R. Divakaran, S. Anand, and S. B. Shenoy. "Advances in the Whipple Shield Design and Development:" *Journal of Dynamic Behavior of Materials* 8, no. 1 (March 1, 2022): 20–38. https://doi.org/10.1007/s40870-021-00314-7.

Paragon Space Development Corporation. "Environmental Control & Life Support Systems (ECLSS)." Accessed October 26, 2022. https://www.paragonsdc.com/what-we-do/life-support/environmental-control-life-support-systems-eclss/.

Petty, John Ira. "Astronauts' Dirty Laundry." Feature Articles. NASA News. Brian Dunbar, April 10, 2003. https://www.nasa.gov/vision/space/livinginspace/Astronaut_Laundry.html.

P&G. "P&G's Tide Is Headed to Space with NASA and SpaceX!," December 20, 2021. https://us.pg.com/blogs/tide-is-headed-to-space/.

Phillips, Tony. "Greenhouses for Mars." Feature Articles. NASA, February 25, 2004. https://www.nasa.gov/vision/earth/livingthings/25feb_greenhouses.html.

Phillips, Tony, and Donald Pettit. "Saturday Morning Science | Science Mission Directorate." NASA Science, February 25, 2003. https://science.nasa.gov/science-news/science-at-nasa/2003/25feb_nosoap.

Pierce, Margo. "A New Doorway to Space." Text. NASA, November 13, 2020. http://www.nasa.gov/directorates/spacetech/spinoff/New_Doorway_to_Space.

Pine II, B. Joseph, and James H. Gilmore. "Welcome to the Experience Economy." *Harvard Business Review*, July 1, 1998. https://hbr.org/1998/07/welcome-to-the-experience-economy.

Pixar Animation Studios. "WALL-E." Accessed November 22, 2022. https://www.pixar.com/feature-films/walle.

"Plants That Clean Water | Kellogg Garden Organics™." Accessed March 21, 2023. https://kellogggarden.com/blog/gardening/plants-that-clean-water/.

Pline, Alex. "NASA—Mossy Space Spirals." Feature Articles. NASA. Brian Dunbar, July 16, 2002. https://www.nasa.gov/vision/earth/livingthings/16jul_firemoss.html.

Podsada, Janice. "Marysville Startup Prepares for Space — the Financial Frontier." HeraldNet.com, September 6, 2023. https://www.heraldnet.com/business/marysville-startup-prepares-for-space-the-financial-frontier/.

PRNewswire. "Robotic Bartender Market Size to Grow by USD 678.8 Mn, Bars and Pubs Leveraging the Use of Advanced Technology Will Drive Growth—Technavio," September 29, 2022. https://www.prnewswire.com/news-releases/robotic-bartender-market-size-to-grow-by-usd-678-8-mn-bars-and-pubs-leveraging-the-use-of-advanced-technology-will-drive-growth---technavio-301635702.html.

Pultarova, Tereza. "Abundant Harvest in Antarctic Greenhouse Shows Promise for Moon Agriculture." Space.com, May 12, 2021. https://www.space.com/mars-lunar-greenhouse-antarctica-harvest.

———. "Could Space Greenhouses Solve Earth's Food Crisis?" Space.com, February 12, 2021. https://www.space.com/space-greenhouses-nanoracks-food-crisis.

Rainey, Kristine, and Linda Herridge. "Crew Members Sample Leafy Greens Grown on Space Station." Text. NASA, August 7, 2015. http://www.nasa.gov/mission_pages/station/research/news/meals_ready_to_eat.

RENERGON—ANAEROBIC DIGESTION. "Anaerobic Digestion Explained," July 25, 2021. https://www.renergon-biogas.com/en/anaerobic-digestion-explained/.

Research, Polaris Market. "Global Cultured Meat Market Size Estimated to Reach USD 499.9 Million By 2030, With 16.2% CAGR: Statistics Report by Polaris Market Research." PR Newswire. Accessed December 2, 2022. https://www.prnewswire.com/news-releases/

global-cultured-meat-market-size-estimated-to-reach-usd-499-9-million-by-2030--with-16-2-cagr-statistics-report-by-polaris-market-research-301523708.html.

ResearchGate. "What Plants Could Be Useful for Greywater Purification?," June 12, 2012. https://www.researchgate.net/post/What-plants-could-be-useful-for-greywater-purification.

Roach, Mary. *Packing for Mars: The Curious Science of Life in the Void.* 1st Edition. New York, NY: W. W. Norton & Company, 2010. https://wwnorton.com/books/9781324036050/about-the-book/description.

Robert Jacobson. *Space Is Open for Business: The Industry That Can Transform Humanity*. 1st Edition. Los Angeles, CA: Self-published, Robert Jacobson, 2020. https://www.robertjacobson.com/book.

Roberts, Thomas. "International Astronaut Database." Aerospace Security, July 5, 2022. https://aerospace.csis.org/data/international-astronaut-database/.

Robinson, Ed, and Carolyn Robinson. *The "Have-More" Plan: A Little Land—A Lot of Living*. Reprint. North Adams, Massachusetts: Storey Publishing, LLC, 1973. https://www.hachettebookgroup.com/titles/ed-robinson/the-have-more-plan/9780882660240/.

"RoboGames! (Formerly ROBOlympics)." Accessed July 13, 2023. http://robogames.net/index.php.

Robot Bartenders of BarBot 2013 Serve Up Drinks. SOMArts Art Gallery: BarBot 2013, 2013. https://www.youtube.com/watch?v=zQaZ_-i0EJM.

Rockoff, L. A., R. F. Raasch, and R. L. Peercy. "Space Station Crew Safety Alternatives Study. Volume 3: Safety Impact of Human Factors," June 1, 1985. https://ntrs.nasa.gov/citations/19850021672.

Rodriguez-Carias, Abner A., John Sager, Valdis Krumins, Richard Strayer, Mary Hummerick, and Michael S. Roberts. "In-Vessel Composting of Simulated Long-Term Missions Space-Related Solid Wastes." NASA John F. Kennedy Space Center, Publication: 2002 Research Reports: NASA/ASEE Fellowship Program, December 1, 2002. https://ntrs.nasa.gov/citations/20030062829.

Roman Strigil. "Collections Database." Accessed March 21, 2023. https://museums.fivecolleges.edu/detail.php?t=objects&type=ext&id_number=AC+1946.94.

Ross, Charlie Bradley. "Fabric Made From Fungi." *The Sustainable Fashion Collective* (blog), August 24, 2016. http://www.the-sustainable-fashion-collective.com/2016/08/24/fabric-made-fungi.

Roulette, Joey. "SpaceX's Toilet Is Working Fine, Thanks for Asking." *The New York Times*, November 11, 2021, sec. Science. https://www.nytimes.com/2021/11/10/science/spacex-toilet-diapers.html.

———. "The Toilet on the Crew Dragon Capsule Was out of Service. The Crew Had to Use Diapers." *The New York Times*, November 9, 2021, sec. Science. https://www.nytimes.com/2021/11/08/science/spacex-diapers-toilet.html.

Sack, Fred. "'Moss In Space' Project Shows How Some Plants Grow Without Gravity." *Ohio State News* (blog), January 25, 2005. https://news.osu.edu/moss-in-space-project-shows-how-some-plants-grow-without-gravity/.

Sands, Kelly. "NASA Glenn Interns Take Space Washing Machine Designs for a Spin." Text. NASA Glenn Research Center, July 28, 2021. http://www.nasa.gov/feature/glenn/2021/nasa-glenn-interns-take-space-washing-machine-designs-for-a-spin.

———. "Stirling Convertor Sets 14-Year Continuous Operation Milestone." NASA, April 16, 2020. http://www.nasa.gov/feature/glenn/2020/stirling-convertor-sets-14-year-continuous-operation-milestone.

Saria, Lauren. "This Restaurant Is Run Entirely By Robots." Eater SF, August 17, 2022. https://sf.eater.com/2022/8/17/23308389/mezli-robot-restaurant-open-menu-san-francisco.

Saturday Morning Science: Drinking Tea with Chopsticks in Microgravity Onboard the International Space Station. International Space Station, 2006. https://www.youtube.com/watch?v=7obLT4s2-HA.

Sayette, Michael A. "Does Drinking Reduce Stress?" *Alcohol Research & Health* 23, no. 4 (1999): 250–55. https://www.ncbi.nlm.nih.gov/pmc/articles/PMC6760384/.

Scharping, Nathaniel. "What Keeps an Astronaut Awake at Night? Cosmic Rays." Discover Magazine, December 19, 2017. https://www.discovermagazine.com/the-sciences/what-keeps-an-astronaut-awake-at-night-cosmic-rays.

———. "What Would Happen If Earth Stopped Spinning?" *Astronomy Magazine* (blog), April 15, 2021. https://www.astronomy.com/science/what-would-happen-if-earth-stopped-spinning/.

Schneider, William C. "Skylab Lessons Learned as Applicable to a Large Space Station, 1967-1974." Washington, D.C.: National Aeronautics and Space Administration, April 1, 1976. https://ntrs.nasa.gov/citations/19760022256.

Schweiger, David Valentin. "Life on Mars—attempt to grow moss in space." *University Post, University of Copenhagen* (blog), June 17, 2015. https://uniavisen.dk/en/life-on-mars-attempt-to-grow-moss-in-space/.

Science. "Did Mars's Magnetic Field Die With a Whimper or a Bang?," April 30, 2009. https://www.science.org/content/article/did-marss-magnetic-field-die-whimper-or-bang.

ScienceCast 234: The Power of Light. NASA Image and Video Library: NASA HQ, 2016. https://images.nasa.gov/details-234_PowerOfLight.

Scoville Scale. "Chili Pepper Scoville Scale | Scovillescale.Org." Accessed November 27, 2023. https://scovillescale.org/chili-pepper-scoville-scale/.

Seal, Laura. "Mumm to Launch Space Champagne for Astronauts." Decanter, June 8, 2018. https://www.decanter.com/wine-news/mumm-launch-space-champagne-astronauts-395136/.

Shelar, A B, Shradha M Kalburgi, Neha D Kesare, Mr Santosh U Kushwah, and Mr Sagar J Choudhari. "Research Paper on Treatment of Grey Water Using Low Cost Technology For Kushvarta Kund Water" 06, no. 05 (May 2019). https://www.irjet.net/archives/V6/i5/IRJET-V6I51134.pdf.

Shuttle's Toilet Requires Special Training. Vol. 1. STS-132: Behind the Scenes. NASA YouTube Channel, 2010. https://www.youtube.com/watch?v=m1wwzwvfsC0.

Silverman, Jason, Andrew Irby, and Theodore Agerton. "Development of the Crew Dragon ECLSS," ICES-2020-333:11 pages. 2020 International Conference on Environmental Systems, 2020. https://ttu-ir.tdl.org/bitstream/handle/2346/86364/ICES-2020-333.pdf.

"Simply Hydroponics—Replace Bulb." Accessed March 21, 2023. https://www.simplyhydro.com/system/.

Space Frontier Foundation. "Space Frontier Foundation." Accessed July 13, 2023. https://spacefrontier.org/.

Space Games Federation. "Space Games Federation—Sports in Space #EqualSpace™," 2022. https://spacegamesfederation.com/.

Space Station Astronauts Grow a Water Bubble in Space. NASA's Marshall Space Flight Center, 2014. https://www.youtube.com/watch?v=9ZEdApyi9Vw.

Space Studies Institute. "Reconsidering Artificial Gravity for 21st Century Space Habitats by SSI SA Peter Diamandis." Accessed November 28, 2023. https://ssi.org/programs/ssi-g-lab-project/artificial-gravity-peter-diamandis-1987/.

Space Tourism Society. "Space Tourism Society | Welcome." Accessed December 18, 2022. https://spacetourismsociety.org/.

SPACEARCHITECT.ORG. "Space Architect," May 17, 2019. https://spacearchitect.org/.

Spencer, John, and Karen Rugg. *Space Tourism: Do You Want to Go?* 1st Edition. Burlington, Ontario, Canada: Apogee Books, 2004. https://www.cgpublishing.com/Books/SpaceTourism.html.

Staff, ScienceAlert. "A Chunk of Satellite Almost Hit The ISS, Requiring an 'Urgent Change of Orbit.'" ScienceAlert, November 15, 2021. https://www.sciencealert.com/a-chunk-of-chinese-satellite-almost-hit-the-international-space-station.

Stapleton, Thomas J., James L. Broyan, Shelly Baccus, and William Conroy. "Development of a Universal Waste Management System." In *43rd International Conference on Environmental Systems*. Vail, CO: American Institute of Aeronautics and Astronautics, 2013. https://doi.org/10.2514/6.2013-3400.

"Starfish Space." Accessed October 29, 2023. https://www.starfishspace.com/.

Stine, G. Harry. *Halfway to Anywhere: Achieving America's Destiny in Space*. 1st Edition. New York, New York: M. Evans and Company, Inc., 1996.

"Stirling Engine—Energy Education." Accessed October 16, 2022. https://energyeducation.ca/encyclopedia/Stirling_engine.

Story, David. "UA-CEAC Prototype Lunar Greenhouse." University of Arizona, College of Agriculture and Life Sciences, Controlled Environment Agriculture Center, 2010. https://www.ag.arizona.edu/lunargreenhouse/.

Tandy, Bill. "GRAVITICS Overview." Future In-Space Operations (FISO) Telecon Presentations Archive 2022-present, May 10, 2023. https://fiso.spiritastro.net/telecon/Tandy_5-10-23/.

TANGERMANN, VICTOR. "As Spacecraft Toilet Rumors Swirl, Bidet Company Pitches Elon Musk." Futurism, September 23, 2021. https://futurism.com/bidet-company-elon-musk.

TapFlow Pumps UK. "How Does A Peristaltic Pump Work?" Accessed November 7, 2023. https://www.tapflopumps.co.uk/blog/peristaltic-pump-guide/.

Tavares, Frank. "Meet ISAAC, Integrating Robots with the Space Stations of the Future." Text. NASA, August 10, 2021. http://www.nasa.gov/feature/ames/meet-isaac.

Taylor, Zachary. "A Study of Space Bathroom Design." *Acta Astronautica* 174 (September 1, 2020): 55–60. https://doi.org/10.1016/j.actaastro.2020.04.027.

"The Coriolis Effect: Earth's Rotation and Its Effect on Weather." Accessed November 9, 2023. https://education.nationalgeographic.org/resource/coriolis-effect.

The Jetsons Fandom Wiki. "Rosey the Robot." Accessed December 7, 2023. https://thejetsons.fandom.com/wiki/Rosey.

The Power of Light. "The Power of Light | Science Mission Directorate." NASA Science, December 13, 2016. https://science.nasa.gov/news-articles/the-power-of-light.

The Scrubba Wash Bag. "Scrubba Wash Bag—Travel Washing Machine." Accessed November 2, 2022. https://thescrubba.com/products/scrubba-wash-bag.

ThinkOrbital. "ThinkOrbital—We Build Space," October 30, 2023. https://thinkorbital.com/.

This Big Greenhouse Built around a House Is the Home for a Family of 7. Murissonstraat, 8930 Rekkem (Menen), Belgium, 2021. https://www.youtube.com/watch?v=atc6-JCVIOs.

Timmermann, Lisa F., Klaus Ritter, David Hillebrandt, and Thomas Küpper. "Drinking Water Treatment with Ultraviolet Light for Travelers—Evaluation of a Mobile Lightweight System." *Travel Medicine and Infectious Disease* 13, no. 6 (November 1, 2015): 466–74. https://doi.org/10.1016/j.tmaid.2015.10.005.

Trevino, Emma. "Your Softest, Cleanest Skin Yet: The Body Plane Tool." *Esker* (blog), June 17, 2020. https://eskerbeauty.com/products/body-plane.

Turchenko, Alina. "Brand New Dining Experience: Top 5 Automated Restaurants." PaySpace Magazine, March 31, 2022. https://payspacemagazine.com/tech/brand-new-dining-experience-top-5-automated-restaurants/.

"Ultraviolet (UV) Radiation | Center for Science Education." Accessed December 7, 2023. https://scied.ucar.edu/learning-zone/atmosphere/ultraviolet-uv-radiation.

United Nations, Department of Economic and Social Affairs. "World Population to Reach 8 Billion on 15 November 2022." United Nations, November 15, 2022. https://www.un.org/en/desa/world-population-reach-8-billion-15-november-2022.

United States Environmental Protection Agency. "Showerheads." Overviews and Factsheets, October 14, 2016. https://www.epa.gov/watersense/showerheads.

University of Florida, Institute of Food and Agricultural Sciences—UF/IFAS. "Space Plants Lab." Accessed March 21, 2023. https://hos.ifas.ufl.edu/spaceplantslab/.

U.S. Department of Agriculture, Agricultural Research Service. "What Is Pyrolysis?," September 10, 2021. https://www.ars.usda.gov/northeast-area/wyndmoor-pa/eastern-regional-research-center/docs/biomass-pyrolysis-research-1/what-is-pyrolysis/.

UT Southwestern Medical Center Newsroom. "High-Tech Sleeping Bag Could Solve Vision Issues in Space," December 13, 2021. https://www.utsouthwestern.edu/newsroom/articles/year-2021/high-tech-sleeping-bag.html.

"Vallejo Wastewater, CA | Official Website." Accessed May 11, 2023. https://www.vallejowastewater.org/.

Vandewalle, Koen, and Samia Wielfaert. "Kaseco | The Ultimate Greenhouse." Kaseco Plus. Accessed October 26, 2022. https://www.kaseco.plus/en.

Vast Space LLC. "Roadmap — VAST." Accessed January 10, 2024. https://www.vastspace.com/roadmap.

Wade, Mark. "DC-X." Encyclopedia. Encyclopedia Astronautica, December 28, 2012. https://web.archive.org/web/20121228125150/http://www.astronautix.com/lvs/dcx.htm.

Waldie, James M., and Dava J. Newman. "A Gravity Loading Countermeasure Skinsuit." *Acta Astronautica* 68, no. 7 (April 1, 2011): 722–30. https://doi.org/10.1016/j.actaastro.2010.07.022.

WALL-E. Animation. Walt Disney Home Entertainment, 2008.

Warren, Le. "Rubber Vacuum Pants That Suck." NASA Blog: A Lab Aloft (International Space Station Research), June 2, 2015. https://blogs.nasa.gov/ISS_Science_Blog/2015/06/02/rubber-vacuum-pants-that-suck/.

Water Glove in Space, 2013. https://www.youtube.com/watch?v=aD1D9E_tdS8.

"Weightless Washcloths and Floating Showers." Accessed March 21, 2023. https://www.esa.int/Science_Exploration/Human_and_Robotic_Exploration/Business/Weightless_washcloths_and_floating_showers.

Weinersmith, Kelly, and Zach Weinersmith. "A City on Mars." PenguinRandomhouse.com. Accessed January 31, 2024. https://www.penguinrandomhouse.com/books/639449/a-city-on-mars-by-kelly-and-zach-weinersmith/.

Weislogel, M. M., J. C. Graf, A. P. Wollman, C. C. Turner, K. J. T. Cardin, L. J. Torres, J. E. Goodman, and J. C. Buchli. "How Advances in Low-g Plumbing Enable Space Exploration." *Npj Microgravity* 8, no. 1 (May 20, 2022): 1–11. https://doi.org/10.1038/s41526-022-00201-y.

"What Is the Coriolis Effect? | NOAA SciJinks—All About Weather." Accessed November 9, 2023. https://scijinks.gov/coriolis/.

White, Frank. *The Overview Effect: Space Exploration and Human Evolution*. 1st Edition. Boston, Massachusetts: Houghton Mifflin Company, 1987. https://www.weareoverview.com/.

Willberg, Henriette. "Gloios: Grime, Sweat and Olive Oil." *Ancient Anatomies* (blog), November 21, 2017. https://ancientanatomies.wordpress.com/2017/11/21/gloios-grime-sweat-and-olive-oil/.

Wolfrom, Jessica. “When Fashion Is Fungal.” Washington Post, August 31, 2020. https://www.washingtonpost.com/climate-solutions/2020/08/31/fashion-musrhooms-mycelium-climate/.

World Economic Forum. “Houseplants Can Improve Your Mental Health and Wellbeing. Here’s How,” August 5, 2022. https://www.weforum.org/agenda/2022/08/houseplants-nature-mental-health-greenery-cognition/.

World Health Organization. “Food, Genetically Modified.” Accessed November 23, 2022. https://www.who.int/health-topics/food-genetically-modified.

WorldAtlas. “What Is the Coriolis Effect?,” May 1, 2023. https://www.worldatlas.com/oceans/what-is-the-coriolis-effect.html.

Wringing out Water on the ISS—for Science! International Space Station: Canadian Space Agency, 2013. https://www.youtube.com/watch?v=o8TssbmY-GM.

XPRIZE. “XPRIZE Foundation Ansari Prize | XPRIZE Foundation.” Accessed July 13, 2023. https://www.xprize.org/prizes/ansari.

Yamamoto, Minehide, Naoki Nishikawa, Hiroyuki Mayama, Yoshimune Nonomura, Satoshi Yokojima, Shinichiro Nakamura, and Kingo Uchida. “Theoretical Explanation of the Lotus Effect: Superhydrophobic Property Changes by Removal of Nanostructures from the Surface of a Lotus Leaf.” *Langmuir: The ACS Journal of Surfaces and Colloids.* 31, no. 26 (July 7, 2015): 7355–63. https://doi.org/10.1021/acs.langmuir.5b00670.

Yoga in Space 2—Jaime Visits the European Astronaut Centre I Cosmic Kids Special Project, 2022. https://www.youtube.com/watch?v=3Npx_ZG95qM.

Yuri’s Night. “Yuri’s Night—The World Space Party.” Accessed July 13, 2023. https://yurisnight.net/.

Zehnder, Jeff. “New FRIDGE Could Bring Real Ice Cream to Space.” Bioserve Space Technologies, University of Colorado Boulder, Ann and H.J. Smead Aerospace Engineering Sciences, College of Engineering and Applied Science, April 23, 2020. https://www.colorado.edu/aerospace/2020/04/23/new-fridge-could-bring-real-ice-cream-space.

Zer Era. “ZER Era (3D Printed Fashion House).” Accessed December 11, 2022. https://zereraofficial.com/collections/all.

“Zimpro Wet Air Oxidation Process | Lummus Technology.” Accessed January 14, 2024. https://www.lummustechnology.com/sustainability/green-circle/water-and-wastewater-technologies/zimpro-wet-air-oxidation-process.

“Zimpro® Wet Air Oxidation.” Accessed January 14, 2024. https://www.wateronline.com/doc/zimpro-wet-oxidation-0001.

“Особенности Космической Бани: Космонавты Тоже Любят Попариться,” March 18, 2019. https://9ban.ru/vidi/v-kosmicheskoj-bane.

About the Author

Samuel Coniglio is a futurist, technical writer, photographer, inventor, and private space industry advocate. He is a Board member and former Vice President of the Space Tourism Society. His research and presentations have been covered by international media, including the Discovery Channel and History Channel. One of his more famous concepts, the Zero Gravity Cocktail Glass Project, won international acclaim and reinvented the concept of how to drink liquids in a microgravity environment. His cocktail making-robots have been showcased on the Discovery Channel and made the cover of the Wall Street Journal.

Samuel was active with the National Space Society, Space Frontier Foundation, and Yuri's Night World Space Party. In the 1990's, he worked for McDonnell Douglas (now Boeing) on the Space Shuttle, International Space Station, and the little-known, but historic Delta Clipper Experimental (DC-X, DC-X/A) reusable rocket program. In 2004, he helped run logistics for the XPRIZE Team film crew during the historic flights of SpaceShipOne.

As a regular at San Francisco Bay Area Maker Faires, Samuel worked as crew for the Steampunk art car studio known as Obtainium Works, where they built and drove dozens of whimsical fantasy vehicles for parades, shows, and Burning Man. For the Robogames hobby robotics event, Samuel built, or collaborated in building, three cocktail-making robots (drinkbots), which manage to dispense drinks to customers in hilarious and entertaining ways. Samuel also participates at the Burning Man event, where he assists artists building unique kinetic art pieces, and as a Black Rock Ranger: a volunteer first responder to assist people in need, mediate disputes, and be a liaison with police, fire, and medical teams. Using the harsh Black Rock Desert in Nevada as a space analog, he learned that the key to human survival anywhere is building lots of physical and logistical infrastructure, and also being kind to people.

Since 1999, Samuel has been researching and prototyping off-world domestic concepts with the Space Tourism Society, and making presentations about potential designs to make life easier "for the rest of us" in space. This book is the culmination of years of experience in these broad fields.